Oliver Niggemann · Peter Schüller
Editors

IMPROVE – Innovative Modelling Approaches for Production Systems to Raise Validatable Efficiency

Intelligent Methods for the Factory of the Future

Editors
Oliver Niggemann
inIT – Institut für industrielle Informationstechnik
Hochschule Ostwestfalen-Lippe
Lemgo, Germany

Peter Schüller
Institut für Logic and Computation
Vienna University of Technology
Wien, Austria

https://doi.org/10.1007/978-3-662-57805-6

Preface

Within the manufacturing industry, the complexity of production plants is steadily increasing due to more product variances, product complexity, and pressures for production efficiency. Production systems must operate more autonomously, creating challenges for larger Industries and serious problems for SMEs without the needed expertise or sufficient resources to adapt new technical possibilities.

To meet these challenges, the European research and innovation project IMPROVE — Innovative Modelling Approaches for Production Systems to Raise Validatable Efficiency — has developed novel data-based solutions to enhance machine reliability and efficiency. Innovative solutions in the fields of artificial intelligence, simulation & optimization, condition monitoring and alarm management provide manufacturers with a human machine interface (HMI) and decision support system (DSS) to ensure an optimized production.

This book presents a selection of results from the IMPROVE project. Methods from artificial intelligence, software engineering, social science, and machine learning are employed for solving important Industry 4.0 topics such as self-optimization, anomaly detection, alarm flood management, and root cause analysis.

The editors would like to thank all authors which provided high-quality contributions and all reviewers which spent significant effort to make each of the following chapters scientifically sound as well as practically relevant in an industrial setting. We hope that this collection will form a valuable addition to the knowledge in the research fields of Machine Learning, technologies for Cyber-Physical Systems and Industry 4.0.

The IMPROVE project has received funding from the European Union's Horizon 2020 research and innovation programme under grant agreement No. 678867.

Prof. Dr.-Ing. Oliver Niggemann
Dr. Dipl.-Ing. Peter Schüller

Table of Contents

Pages

Concept and Implementation of a Software Architecture for Unifying Data Transfer in Automated Production Systems 1–17
Emanuel Trunzer, Simon Lötzerich and Birgit Vogel-Heuser

Social Science Contributions to Engineering Projects: Looking Beyond Explicit Knowledge Through the Lenses of Social Theory 19–36
Peter Müller and Jan-Hendrik Passoth

Enable learning of Hybrid Timed Automata in Absence of Discrete Events through Self-Organizing Maps 37–54
Alexander von Birgelen and Oliver Niggemann

Anomaly Detection and Localization for Cyber-Physical Production Systems with Self-Organizing Maps 55–71
Alexander von Birgelen and Oliver Niggemann

A Sampling-Based Method for Robust and Efficient Fault Detection in Industrial Automation Processes 73–91
Stefan Windmann and Oliver Niggemann

Validation of similarity measures for industrial alarm flood analysis 93–109
Marta Fullen, Peter Schüller and Oliver Niggemann

Concept for Alarm Flood Reduction with Bayesian Networks by Identifying the Root Cause 111–129
Paul Wunderlich and Oliver Niggemann

Concept and Implementation of a Software Architecture for Unifying Data Transfer in Automated Production Systems

Utilization of Industrie 4.0 Technologies for Simplifying Data Access

Emanuel Trunzer[0000-0002-4319-9801], Simon Lötzerich and Birgit Vogel-Heuser[0000-0003-2785-8819]

Institute of Automation and Information Systems
Technical University of Munich, Munich, Germany
{emanuel.trunzer, simon.loetzerich, vogel-heuser}@tum.de

Abstract. The integration of smart devices into the production process results in the emergence of cyber-physical production systems (CPPSs) that are a key part of Industrie 4.0. Various sensors, actuators, Programmable Logic Controllers (PLCs), Manufacturing Execution Systems (MES) and Enterprise Resource Planning (ERP) systems produce huge amounts of data and meta data that can hardly be handled by conventional analytic methods. The main goal of this work is to develop an innovative architecture for handling big data from various heterogeneous sources within an automated production system (aPS). Moreover, enabling data analysis to gain a better understanding of the whole process, spotting possible defects in advance and increasing the overall equipment effectiveness (OEE), is in focus. This new architecture vertically connects the production lines to the analysts by using a generic data format for dealing with various types of data. The presented model is applied prototypically to a lab-scale production unit. Based on a message broker, the presented prototype is able to process messages from different sources, using e.g. OPC UA and MQTT protocols, storing them in a database and providing them for live-analysis. Furthermore, data can be anonymized, depending on granted access rights, and can be provided to external analyzers. The prototypical implementation of the architecture is able to operate in a heterogeneous environment supporting many platforms. The prototype is stress tested with different workloads showing hardly any response in the form of longer delivery times. Thus, feasibility of the architecture and its suitability for industrial, near real-time applications can be shown on a lab-scale.

Keywords: Automated Production System (aPS), Big Data Applications, Cyber-physical Systems (CPS), Data Acquisition, Data Analysis, Heterogeneous Networks, Industrie 4.0, Industry 4.0, Internet of Things (IoT), Message-oriented Middleware, Systems Architecture

O. Niggemann and P. Schüller (Eds.), *IMPROVE – Innovative Modelling Approaches for Production Systems to Raise Validatable Efficiency*, Technologien für die intelligente Automation 8, https://doi.org/10.1007/978-3-662-57805-6_1

1 Introduction and Motivation

Globalization and high competitive pressure require manufacturing companies to develop new solutions such as the digitalization of existing production processes, massive information exchange, and development of new business models. Embracing new technologies, those efforts are known, amongst others, as Industrie 4.0, Cyber Physical Production Systems (CPPS), or Industrial Internet of Things (IIoT). [1]

A major requirement for leveraging the full potential of Industrie 4.0 applications is the utilization of big data and data analytics methods in production systems. These methods are used to reveal otherwise unknown knowledge, enable process improvements, and increase the overall equipment efficiency (OEE). In modern automated production systems, the generated data shares many similarities with big data, defined via the four V's (volume, variety, velocity, and value) [2]. Several factors handicap automated data analysis in the field of automation. Especially the multitude and heterogeneity of data sources, formats, and protocols due to long lifecycles (up to 30 years) in the production environment pose a challenge (variety). In addition, large amounts of historic data (volume) have to be combined with constantly streamed data from the plant in order to make decisions (value) based on the analysis results in time (velocity). Classical data analysis approaches are not applicable in this heterogeneous automation context. Therefore, new, innovative system architectures for applying big data techniques in automated production systems have to be developed. [3–5]

These difficulties become evident using an example from process industry: A multitude of sensors continuously collect process data, which is stored in databases mainly for documentation purposes. A manufacturing execution system (MES) is used to manage data concerning resource planning and order execution. Moreover, a shift book contains information about operators responsible for surveying the mode of operation and incidents happening during their shifts. Adding additional complexity, quality and maintenance data may be stored in other systems or databases. Together, they form a complex network of interwoven IT systems, based in different physical locations, relying on different, often incompatible data formats. Extracting knowledge from this heterogeneous setup is difficult and can often impossible without a huge manual effort carried out by experts. Consequently, an architecture for unifying data access could greatly enhance the possibilities and impact of data analysis in production environments. This could be achieved by including all relevant sources and providing their data for analysis tools and enabling ubiquitous computing. In this context, we consider the term "architecture" to be defining the description of the overall system layout based on principles and rules in order to describe its construction, enhancement and usage. This definition is compliant with the Reference Architecture Model Industrie 4.0 (RAMI 4.0) [6].

We therefore suggest that the implementation of such a software architecture in production systems is desirable and possible, simplifying data acquisition, aggregation, integration and warehousing, as well as providing data for analysis. This paper describes a generic architecture that can be applied to various scenarios and shows its concrete use and practicability in a lab-scale production system. It pays special attention to the multitude of requirements arising from automated production systems, legacy systems, heterogeneous sources, and data processing,

This contribution is an extended and adapted version of the contribution presented at the 2017 IEEE International Conference on Industrial Technology (ICIT 2017) [7]. In addition to the original version, the literature review is expanded and a prototypical implementation is added.

The remainder of the paper is structured as follows: We first derive requirements for a system architecture to support Industrie 4.0 principles, then evaluate how other authors fulfill them, and identify a research gap. After deriving a concept for a new architecture, we evaluate it, using expert interviews and a prototypical lab-scale implementation. The last part summarizes the findings and provides an outlook for further applications and fields of research.

2 Requirements for a System Architecture to Support Industrie 4.0 Principles

One of the main goals of Industrie 4.0 is the optimization of the manufacturing process based on algorithms and data to increase the OEE [3]. This can be achieved by gaining a better understanding of complex procedures inside the plant and thus reducing maintenance and downtimes. Therefore, new ways of knowledge discovery have to be provided and the vertical integration of the production process has to be enhanced. In order to establish a new system architecture to support Industrie 4.0 principles, several requirements have to be considered.

Supporting various data sources, including legacy systems, is one of the key aspects for successfully implementing a common architecture (requirement R1). Characteristic for the data landscape of a manufacturing enterprise is its heterogeneity and variety of systems. Currently, each layer of an enterprise typically operates using a multitude of layer-specific protocols and formats. These communication channels were often not explicitly designed to operate with other tools. Long lifespans in automation industry ensure the presence of legacy devices that prohibit disruptive changes. [8–10] The new architecture thus has to be able to operate without interfering with the mode of operation of existing systems. Moreover, integration of tools and machines has to be possible regardless of their technology or capabilities [11–13]. A support for various sources can be achieved by defining interfaces, which have to be implemented by data adapters, transferring specific protocols and data formats into a common one.

Thus, we derive as requirement R2 a common data model. As Hashem et al. [2] stated, data staging is an open research issue that needs more attention in the future. Introducing a reference data model [5, 14] can greatly reduce manual work for data integration if it serves towards a common understanding of the data from all involved systems. An exemplary data model is the ontology defined in ISO 15926 Part 8 [15], standardizing data exchange between different IT systems, or the Common Information Model (CIM) defined in the standards EN 61968 and EN 61970 [16, 17] for electrical distribution systems.

Having successfully integrated various sources and transferred data to a common model, dealing with data of different timeliness and message sizes has to be considered. A suitable architecture should be able to process both historic and near real-time data correctly (R3). One the one hand, some sources may constantly send small messages containing information about single variables from within the production process. On the other hand, other sources only send a single message with enormous message size (batch message) when queried, containing for example order information. Some messages may contain near real-time information about current processes, others time-insensitive contents. This implies that the architecture has to be able to extract knowledge from constantly arriving message streams and huge amounts of batch messages. [18] Marz and Warren [19] suggest the so-called lambda architecture. The lambda architecture consists of different layers, namely a speed layer for real-time tasks, a batch layer for accessing historical data from a storage, and a serving layer, combining the results of batch and speed layer. Being capable of handling hybrid processing at near- real-time, their concept is one example for achieving this task.

In order to be applicable for a wide range of use cases, the architecture has to be able to handle various analysis methods and thus provide interfaces for including different GUIs and HMIs, export possibilities, and query tools (R4). In order to simplify data retrieval, the architecture should be able to provide users the collected data at the right level of abstraction (depending on the request) [4]. Also, due to the nature of Big Data, a single form of data visualization is not sufficient. It is problematic to find user-friendly visualizations. Thus, new techniques and frameworks should be includable into the architecture, requiring well-defined interfaces for queries. [20] Begoli and Horey [21] suggest the usage of open, popular standards, exposition of results via web interfaces, and the adoption of lightweight, web-oriented architectures. They also confirm that many different visualizations across many datasets are necessary to extract knowledge from the data.

Not only in-company analysis can be deducted, but to leverage the full potential of the data, a cross-organizational data transfer and analysis process is necessary. This is why as requirement R5 a support for anonymized data transfer across organizational borders is deduced. Manufacturers could develop models for lifetime prediction, improve production lines, and increase the OEE by gathering data from their shipped product. However, once a machine is shipped to the customer, the manufacturers of production machines rarely can access relevant sensor data to predict lifespan and possible failures. Revealing causes for hardware failures can be supported by collecting datasets across corporate boundaries, and analyzing process, device, and servicing data. Relying on more data sources and more data contents, lifespan predictions can become more accurate. Measures for improving the OEE can be deducted through the utilization of better models. [22] In order to guarantee privacy protection and security, those datasets have to be anonymized and transferred via a secure connection to allow analysis by selected personnel. Developing efficient and secure ways for big data exchange is still an open research issue. [2, 23] Protection of trade secrets is a very important aspect for every company, making data transfer across organizational borders a highly sensitive topic. [2]

All derived requirements are summarized in **Table 1**.

Table 1. Table of Requirements for Industrie 4.0 System Architectures.

ID	Description
R1	**Support of various data sources, including legacy systems.** Handling of heterogeneous data from all relevant sources, being able to include existing legacy systems.
R2	**Usage of a common data model.** A common understanding of the aggregated data is necessary to simplify the analysis task. This implies the definition of a common data model and can greatly simplify the communication between different data sources, services and business units.
R3	**Processing of historic and near real-time data.** Being able to handle both batch messages and stream data is an integral requirement for processing messages with different timeliness.
R4	**Support different analysis methods and tools.** Different levels of abstraction and support for various analysis tools are important to support users in detecting patterns among the data.
R5	**Support anonymized data transfer across organizational borders** By including data from various parties into analysis, life span predictions can become more accurate and models for determining possible failures can become more sophisticated. In order to guarantee privacy protection and protection of trade secrets, data transfer has to be done in an anonymized and secure way.

3 State-of-the-Art of Industrie 4.0 System Architectures

Several reference architectures exist in the context of Industrie 4.0 and the Industrial Internet of Things (IIoT). The most important ones are the German Reference Architecture Model Industrie 4.0 (RAMI 4.0) [6], the American Industrial Internet Reference Architecture (IIRA) [24] and the draft international standard ISO/IEC CD 30141 [25] for the Internet of Things Reference Architecture (IoT RA). These reference architectures provide an abstract, technology-neutral representation of an IIoT system and rules for the development of an actual architecture. Therefore, they feature an abstract description, which has to be adopted in order to represent the specific characteristic of an actual system.

The concept of the Enterprise Service Bus (ESB) was proposed by Chappell [26]. The ESB relies on a communication and integration backbone to connect different applications and technologies in an enterprise. It employs web service technologies and supports various communication protocols and services. One of the main goals of the ESB is to include various heterogeneous sources and services (R1). This is achieved by using a common data model (R2) for passing messages over the central bus. The design of the ESB also supports different analysis methods (R4), since existing APIs can be used. Anonymized data transfer across organizational borders (R5) is not covered by the ESB. Different data processing ways (R3) are not explicitly mentioned in the description of the ESB and would require special attention during the design and implementation of an Enterprise Service Bus.

The Automation Service Bus (ASB), introduced by Moser, Modrinyi and Winkler [27] is based on the concept of the ESB, but narrows down the scope to engineering tasks in the field of mechatronics development. The ASB uses a common data model (R2) to provide a platform for all disciplines taking part in the development of mechatronic systems. Since the ASB focuses on engineering phase of the asset lifecycle, operational data is not supported (R1). A special consideration of legacy systems is not mentioned, neither whether data adapters exist. Different processing ways (R3) and anonymized data transfer between enterprises (R5) do not exist in the concept. Since the ASB is tailored towards the developed Engineering-Cockpit as a user interface, other analysis methods (R4) are not supported.

The Namur Open Architecture (NOA) presented by Klettner et al. [28] is an additive structure to the conventional production pyramid [29]. Its structure allows an open information exchange over a secondary communication channel between not-neighboring automation layers and a secure backflow from an IT environment into process control. The Namur Open Architecture specifies how information is transferred from the core process control to plant specific monitoring and optimization applications. This is achieved by using open and vendor-independent interfaces. A special interest of the NOA is to support various existing systems and data sources (R1). The architecture can be connected to various applications and means of analysis (R4). A cross-organizational data transfer is intended via the "Central Monitoring + Optimization" part, however, anonymization is not mentioned (R5). NOA describes two channels for data transfer from shop floor devices to the analysis part ("Central M+O"). The direct path could be used to transfer soft real-time data, whereas open interfaces could be used to process batch data (R3). The NOA relies on a common data model (R2) serving as unified space of information for process data.

The PERFoRM project [30] focuses on seamless production system reconfiguration, combining the human role as flexibility driver and plug-and-produce capabilities. It is compliant with legacy systems by using proper adapters to translate all data into a system-wide language. Special focus is put on integrating systems from all stages of the production pyramid using suitable wrappers, interfaces and adapters (R1). Adapters transfer data models of legacy devices into a common data model (R2). The architecture's middleware ensures reliable, secure and transparent interconnection of the various hardware devices and software application. It therefore follows a service-oriented principle, i.e. offering functionalities as services, which can be discovered and requested by other components. Since it was designed in a human centered way, various HMIs and analysis methods are supported (R4). The proposed middleware is independent from a certain network technology; various devices implementing the defined interface can be used for analyzing results. A way for transferring data between companies and anonymizing them (R5) is not described by the PERFoRM architecture. The different processing ways for batch data and soft real-time data (R3) are not taken into consideration by the concept.

The Line Information System Architecture (LISA), proposed by Theorin et al. [11], is an event-driven architecture featuring loose coupling and a prototype-oriented information model. LISA puts special focus on integration of devices and services on all levels (R1), reducing effort of hardware changes and delivering valuable aggregated information in the form of KPIs. It uses simple messages from various kinds of devices that are translated via data adapters to fit the LISA message format (R2) and then sent to an Enterprise Service Bus (ESB). The ESB offers publish-subscribe functionality and forwards the messages to all interested receivers. Knowledge extracted from the messages can be used for online monitoring, control, optimization, and reconfiguration. By using data adapters, various analysis methods can be integrated (R4). As not being based on point-to-point communication, the event-driven architecture can be easily reconfigured. Since new event creators only need to know that the event occurred, but not who is receiving the message or how it has to be processed, loose-coupling is achieved. Real-time data will be available directly from the events and historic data can be queried from a suitable storage system (R3). Data exchange across organizational borders is not intended by the architecture (R5).

Hufnagel and Vogel-Heuser [31] present a concept for facilitating the capturing of distributed data sources, relying on the ESB-principle. Their model-based architecture uses data mapping and adapters to transfer data of various sources (R1) into a common data model (R2). It is able to handle batch data and real-time data from different systems (R3). However, the concept does not focus on analysis (R4), or sharing and anonymizing data (R5).

The presented concepts and the degree of requirement fulfillment are summarized in **Table 2.** None of the existing approaches fulfills all derived requirements.

Table 2. Classification Matrix for Existing Industrie 4.0 System Architecture Concepts.

Concept	R1	R2	R3	R4	R5	Implemented
ESB [26]	+	+	O	+	O	+
ASB [27]	-	+	-	-	-	+
NOA [28]	+	+	O	+	O	+
PERFoRM [30]	+	+	-	+	-	-
LISA [11]	+	+	+	+	-	+
Hufnagel and Vogel-Heuser [31]	+	+	+	-	-	-

Legend: + fulfilled, O partly fulfilled, - not fulfilled

4 Concept of a Unified Data Transfer Architecture (UDaTA) in Automated Production Systems

In the following, the concept of the Unified Data Transfer Architecture (UDaTA) will be derived from the aforementioned requirements, paying special attention to its suitability for different use cases. Focus is put on defining the overall concept technology-independently, meaning the specific technologies for an implementation can be adjusted to fulfill the requirements of a given use-cases (e.g. usage of MQTT instead of OPC UA or Kafka [32] instead of an enterprise service bus).

In order to support different analysis methods, various tools (R4) and legacy devices (R1), standardized interfaces are necessary. Relying on a layered structure with well-defined interfaces simplifies reconfiguration and adoption to a variety of use cases. The architecture differentiates between layers for providing raw data, analyzing data, and displaying data. Connecting data sources with its destinations, the so-called *Data Management and Integration Broker* transfers and routes data between the components and layers of UDaTA. This setting allows for both the connection of existing legacy applications as well as newly added tools through interfaces.

Requirement 3 demands for the ability to process both historical and near real-time data. Therefore, UDaTA features a central data storage for saving data and providing it for later analysis. Real time data from data sources is streamed by the broker to the data storage and made available there. Depending on the use-case (e.g. number of sources and message intensity), the data storage can be a relational or non-relational database. Components of the analysis or displaying layer can request data from the storage via the broker, using the aforementioned standard interfaces. Handling all types of data, the central data storage ensures a wide availability of data for all layers. Making not only historic data, but also real-time data available, the Data Management and Integration Broker can stream live data to subscribers of the upper layers, as described by the lambda architecture [19]. This can be important for live-monitoring and optimization operations executed by the analysis layer.

For enabling data transfer across organizational borders (R5), an access control and anonymization component is necessary. Especially, when working with data from other organizational units or companies, the privacy and integrity of data become very important issues. Leaking data, which was not supposed to be transferred, must be avoided. Hence, the broker features an access control and anonymization layer, guaranteeing only approved data access. Before transferring content, data may be anonymized or access restricted, depending on the requesting block and its security clearance.

A common data model (R2) is essential in order to describe the represented system and its data in a unified way and in a language that is understood by all layers and components. The data model of UDaTA has to include representations of the raw data, additional metadata (enriching raw data with information about units of measurement, associated devices, and so on), previously learned and preconfigured models, operator knowledge, parameter sets, and configurations of the service blocks. Each block can work with a subset of the overall data model for carrying out its operations. Data that does not match the common model has to be transferred by adapters in order to be compliant with the other data as presented in [33, 34]. Wrappers can be used to encapsulate third-party applications and provide standardized, compatible interfaces. Those mappings and adaptions have to be carried out by hand, when dealing with a new data model or changes during the asset lifecycle. Especially for legacy blocks, the effort for translating data can sometimes be high, but the benefits of dealing with only one common data format, like easy plug-in of new blocks, as well as high compatibility and increased flexibility, are significant. This is also a reason why a parallel rollout of UDaTA to existing systems and a stepwise adaption is suggested as deployment strategy.

Fig. 1. gives a representation of the Unified Data Transfer Architecture, reflecting one company or organizational structure. Each company or structure can have its own slice of the architecture deployed, allowing a communication over the broker and the analysis of data stored in a different location. Several instances of the broker can communicate with each other and exchange data.

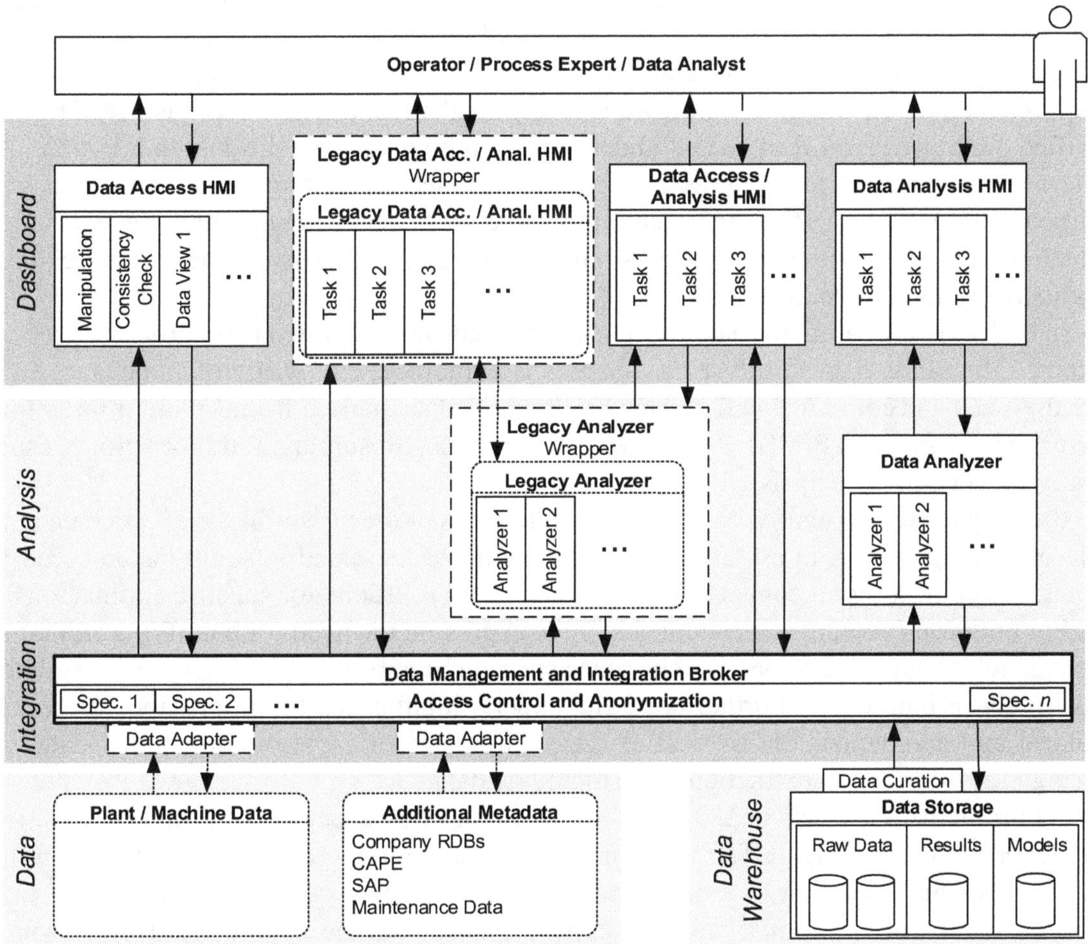

Fig. 1. Schematic Structure of the Unified Data Transfer Architecture (UDaTA).

5 Evaluation

The proposed architecture was evaluated in two ways. At first, an expert evaluation was carried out. Therefore, technical experts were interviewed about an adapted version of the architecture for their specific use-cases. The use-cases originate from different fields, namely process industry and discrete manufacturing. In addition, questionnaires were used to query specific aspects of the architecture.

Furthermore, a prototypical, lab-scale demonstration was implemented using concrete technologies in order to reflect the realization for a specific use-case.

5.1 Expert Evaluation

Semi-structured interviews with technical experts were carried out for two distinct use-cases from the field of process industry and discrete manufacturing. Both use-cases are characterized by a multitude of existing legacy systems, complex system layouts and interactions between software components, as well as a high degree of heterogeneity in the used communication channels (e.g. different field buses) and data formats. For the interviews, the generic UDaTA was adapted for the specific use-cases, reflecting the specific requirements of the field of application. It was evaluated whether there is a need for a unified architecture for data access under the given boundary conditions. In addition, the interviewees were asked about the shortcomings of their current system architectures and if the proposed UDaTA has the potential to solve these disadvantages and inefficiencies. The interviews were supported by questionnaires for capturing specific needs and requirements in the field of application.

In both existing solutions, data is often aggregated and integrated manually by a process expert. Therefore, data from various sources is combined, e.g. written maintenance logs are digitized and combined with the historic process data queried from a plant information management system (PIMS). This task is highly time-consuming and error-prone, especially for repetitive analysis.

Existing legacy systems with proprietary interfaces or the possibility to export data only file-based, e.g. export to csv-files, further complicate the existing solutions and lead to countless point-to-point connections and tailor-made solutions for specific applications.

For both use-cases, the interviewed experts expressed the need for a unifying and innovative architecture. Therefore, the characteristics and flexibility of UDaTA were evaluated positively in both cases. Furthermore, the experts highlighted the importance of access control and anonymization, as well as the significance of cross-organizational data exchange for information extraction from local raw data.

As problematic aspects of UDaTA the definition of an accepted common information model as well as the high effort for a first implementation were mentioned. Overcoming the issue of the definition of a data model is challenging. It can be carried out use-case specific by an interdisciplinary team of experts focusing on the specific needs for the use-case. Alternatively, it could be promoted by a standardization body, for instance following the example of the common information model (CIM) in the energy distribution sector [16, 17]. For the field of automated and cyber-physical production systems, there is an immediate need for such an accepted meta-model. The CIM reflects the evolution of an accepted meta-model unifying the data understanding between different parties and should act as a guidance for such a model in the field of automation.

In order to overcome the complex introduction phase of UDaTA, a stepwise migration is proposed, leaving the existing ISA 95 layout intact. This allows to deploy the architecture in parallel to the existing infrastructure and to integrate the different processes of the production plant over time, therefore benefiting from already migrated applications while leaving the vital, production-relevant infrastructure untouched.

5.2 Prototypical Lab-Scale Implementation

In order to verify the practicability of the concept, a prototypical implementation is deployed on a lab-scale. To simulate a heterogeneous production environment, a bench-scale platform, called extended pick and place unit (xPPU) [35], is used as the main data source for production data. The xPPU is able to sort, pick, and place work pieces, using amongst others cranes, conveyor belts, and a multitude of sensors. As secondary data source, exemplary order data from a csv file is read. The data is sent to a message broker, translated into a common data format and stored in a relational database. Analyzers can send requests for batch data to the broker. These requests are forwarded to the database, the results are anonymized if necessary, and provided to analyzers. In addition, analyzers can subscribe to live data originating from the sources. In order to replicate the heterogeneity of the production environment, various different operating systems, as well as programming languages to realize the separate applications, are used. The whole setup is depicted in **Fig. 2**, showing hardware components, operating systems, software, and exemplary data flows among the elements of the architecture. The setup is described in detail in the following paragraphs.

In the prototypical setup, the bus couplers of the xPPU are communicating via EtherCAT with a Beckhoff CX9020 PLC that runs on Windows 7 CE and TwinCAT 3. This PLC makes its process data available over two different communication channels, namely MQTT messages and an OPC-UA server.

MQTT [36] is a publish-subscribe-based, lightweight messaging protocol that is suitable for Machine-to-Machine (M2M). Using the TwinCAT function block "Tc3_IotBase", MQTT-client functionality becomes available on the PLC. This way, messages with topics like "EnergyMonitoringHardware/CurrentPressure/Int" or "LightGrid/EmergencyStop/Bool" are sent every cycle (currently 10ms). These messages contain the values of the respective variables of the xPPU.

Furthermore, the PLC runs an OPC UA server that provides the states of other selected variables to clients. In our case, a Raspberry Pi 3 is used as a client for the PLC. It operates with Raspbian running an OPC UA client using the .Net Standard reference implementation by the OPC Foundation [37]. Furthermore, it subscribes to a number of variables and translates the received data into the common data model. This data is then sent to the connected broker.

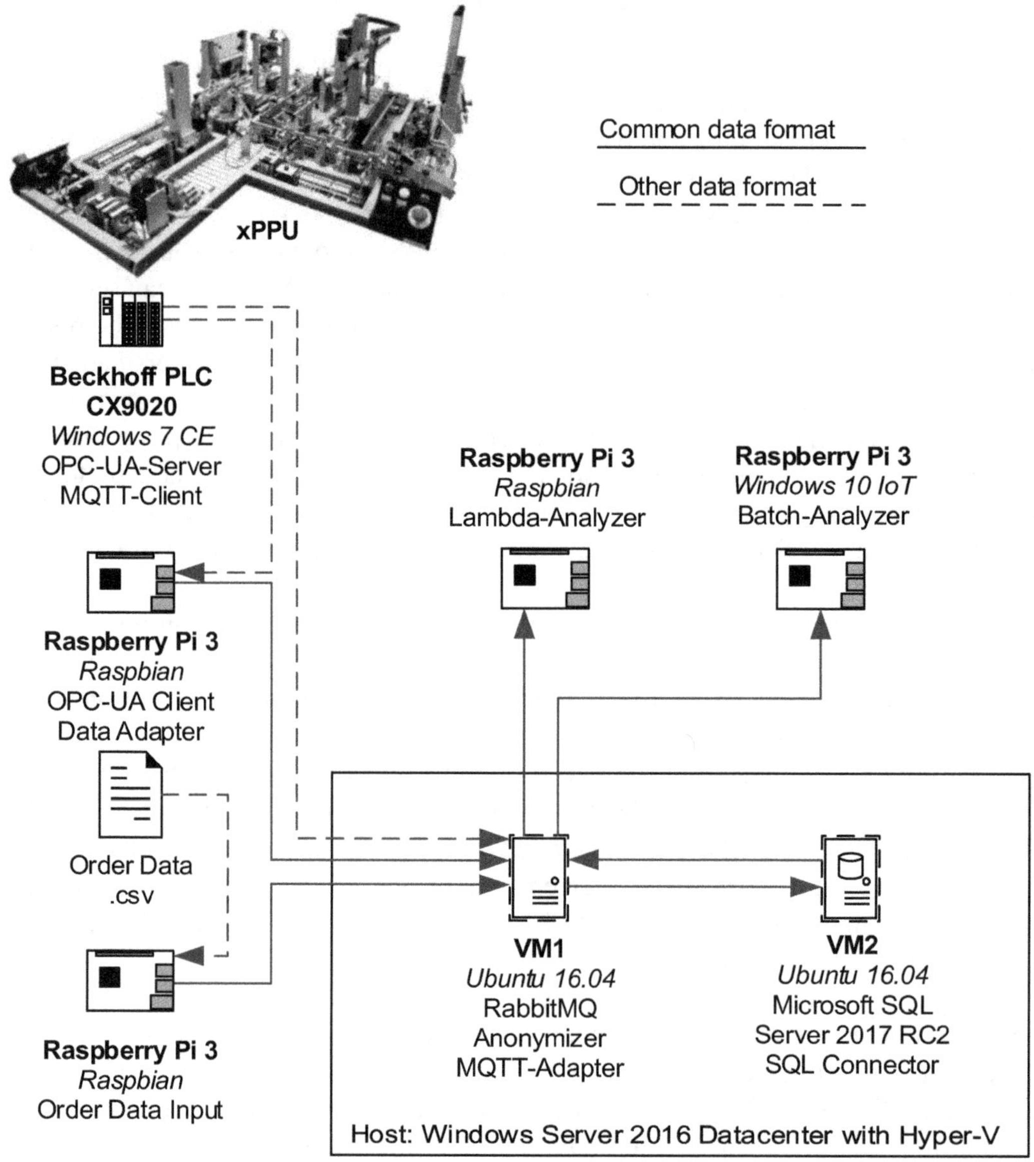

Fig. 2. Hardware and Software Setup of the Prototypical Implementation for the xPPU.

To simulate the processing of batch data, a csv-file containing hypothetic order information is read by a Java program on another Raspberry Pi 3 with Raspbian, translated directly into the common data format and sent to the broker.

Using Hyper-V on a Windows Server 2016 Datacenter host, that is equipped with a Core i7-6700 CPU and 16 GB RAM, we installed Linux Ubuntu 16.04 LTS x64 on two virtual machines (VM).

On the first VM, we installed the open source broker RabbitMQ [38] in version 3.6.11, as well as a .Net Core 2.0 based Anonymizer and Translator for MQTT messages. RabbitMQ works with exchanges, to which messages can be sent. An exchange can be connected to several queues that store messages until they are polled or being subscribed to.

This setup allows for a distribution of messages to the correct destinations. The Translator receives MQTT messages sent by the PLC and translates those into the common data format. If the requesting analyzer has only limited access rights, the Anonymizer for instance changes data and time values to a relative scale in order to minimize information leakage.

The second VM runs the database components. In is this case they compromise a Microsoft SQL Server 2017 RC2 and a .NET Core 2.0 based SQL Connector that receives messages and queries from the broker and handles database communication using Microsoft Entity Framework Core 2.0.

Using two more Raspberry Pis 3, one with Raspbian, one with Windows 10 IoT, instances of .NET Core 2.0 based analyzers are executed. These can subscribe to live data from sources, request data from the database, perform calculations on the data, send calculated data to the broker, or listen to the results of other analyzers.

With this setup, the feasibility of the implementation of an architecture for unifying data transfer in automated production systems is demonstrated. Relying on platform independent technologies like .NET Core 2.0, Java and an open source message broker, that can be executed on Windows, Linux, and macOS, the implementation can be rolled-out in heterogeneous IT environments (R1). By using adapters and translators, the transformation of messages into a common data model is carried out (R2). It is possible to connect arbitrary analyzers (R4) (including legacy components) to the open source message broker RabbitMQ, since it provides clients and developer tools for many programming languages (such as Java, .Net, Ruby, Python, PHP, JavaScript, Objective-C, C, C++), which was also demonstrated in the heterogeneous environment chosen for the demonstration. Analyzers are able to access both historic and live data (R3). Using different roles with different access rights on the broker, data security is ensured. Moreover, as data can be automatically anonymized if necessary, also sensitive information can be shared. Connecting brokers at different physical locations is possible with the Shovel plugin [39] of RabbitMQ (R5). For the realization of UDaTA, it is emphasized, that the selected technologies, languages, or brokers are only of subordinate relevance; the shown prototypical implementation is only one possible solution for this specific use-case.

In order to validate the feasibility of the broker for a large number of incoming and outgoing messages, a time measurement of the delivery times from start to destination is performed. Before carrying out the tests, the internal clocks of each device were synchronized. Afterwards, timestamps are added to the message headers and the timespan it took to deliver the message is calculated. Performing these tests with low message intensity (~2 msg/ sec published), middle intensity (~80 msg/ sec published) and high intensity (~1650 msg/ sec published) for several minutes, no influence of the message intensity on the delivery times is observed. The average message times varied for all intensities between 2.5 and 10.0 milliseconds. These absolute times, however, have to be interpreted with caution since it cannot be guaranteed that the clocks were perfectly synced in the range of milliseconds. What can however be stated is that no evidence of prolonged message times due to higher message intensity is measured for the given setup. Therefore, RabbitMQ is a suitable message broker for providing live data to analyzers in a Unified Data Transfer Architecture with near real-time requirements under the given load scenario.

6 Conclusion and Outlook

Applying big data techniques in the field of manufacturing is currently strongly handicapped due the multitude of protocols and data formats used by legacy devices deployed to the field. Manual data acquisition from closed, proprietary system and subsequent integration of data by experts are often the only chance to access the vast amount of measured data. However, there is a need for making this data automatically available in order to gain information from it, especially with the rise of the ideas of the industrial internet of things (IIoT) and Industrie 4.0. For Cyber-physical systems (CPS) and cyber-physical production systems (CPPS), the transparency of information, as well as big data analytics play a major role. New, flexible architectures are required in order to make this information accessible and apply big data in the field of automation.

This contribution presents a conceptual architecture named UDaTA for data acquisition, integration, and handling from the field device layer up to business applications. Therefore, it provides mechanisms for vertical, as well as horizontal integration. A strong effort is put on the unification of data access and transport for abstracting the complexity of the involved systems. Reference models for IIoT, like the German RAMI4.0 or American IIRA, lay the cornerstones for such an architecture, but due to their generic nature do not capture application specific aspects. Existing concepts for data acquisition and integration in the field of automation often lack the consideration of cross-company data exchange for collaboration and openness of the interfaces. The conceptualized architecture uses a middleware-approach to make the data available and minimizes the number of data transformations between the involved systems by proposing the use of a common information model (CIM).

The concept was evaluated with expert interviews showing an apparent need for the implementation of such an architecture. Moreover, using a bench-scale production system, it was possible to show that UDaTA can be prototypically implemented and is able to transfer and handle data from heterogeneous sources. Acting as middleware component, the open source message broker RabbitMQ received data input from MQTT and OPC UA sources, which was transferred via adapters into a common information model. Both, access to streamed live data and historic batch data, could be demonstrated and opened new possibilities for data analysis.

In order to prove its suitability for real production environments, further testing with larger prototypes and more data sources is required. Special focus has to be put on the formulation of a CIM that is suitable for generic use cases and allows translation from various data formats. Other technologies for the middleware, for example Apache Kafka, OPC UA or DDS-Systems, should also be evaluated and their eligibility for real world use cases compared. Even more important, real analysis of the data and usage of the newly gained knowledge is required. Next steps should focus on using the acquired data for the analysis process.

Acknowledgment

This work is part of the two projects IMPROVE and SIDAP.

IMPROVE (improve-vfof.eu) has received funding from the European Union's Horizon 2020 research and innovation program under grant agreement No 678867.

In addition, SIDAP (www.sidap.de) has received funding by the German Federal Ministry for Economic Affairs and Energy (BMWi) under the grant number 01MD15009F.

References

1. Bauer, H., Baur, C., Camplone, G.: Industry 4.0. How to Navigate Digitization of the Manufacturing Sector. tech. rep., McKinsey Digital (2015)
2. Hashem, I.A.T., Yaqoob, I., Anuar, N.B., Mokhtar, S., Gani, A., Ullah Khan, S.: The Rise of "Big Data" on Cloud Computing. Review and Open Research Issues. Information Systems 47, 98–115 (2015)
3. Vogel-Heuser, B., Hess, D.: Guest Editorial Industry 4.0–Prerequisites and Visions. IEEE Trans. Automat. Sci. Eng. 13, 411–413 (2016)
4. Cecchinel, C., Jimenez, M., Mosser, S., Riveill, M.: An Architecture to Support the Collection of Big Data in the Internet of Things. In: Zhang, L.-J. (ed.) IEEE World Congress on Services (SERVICES), 2014, pp. 442–449 (2014)
5. Jirkovsky, V., Obitko, M., Marik, V.: Understanding Data Heterogeneity in the Context of Cyber-physical Systems Integration. IEEE Trans. Ind. Inf. 13, 660–667 (2017)
6. Deutsches Institut für Normung e.V. (DIN): Reference Architecture Model Industrie 4.0 (RAMI4.0) (2016)
7. Trunzer, E., Kirchen, I., Folmer, J., Koltun, G., Vogel-Heuser, B.: A Flexible Architecture for Data Mining From Heterogeneous Data Sources in Automated Production Systems. In: 2017 IEEE International Conference on Industrial Technology (ICIT), pp. 1106–1111 (2017)
8. Delsing, J., Eliasson, J., Kyusakov, R., Colombo, A.W., Jammes, F., Nessaether, J., Karnouskos, S., Diedrich, C.: A Migration Approach Towards a SOA-based Next Generation Process Control and Monitoring. In: IECON 2011 - 37th Annual Conference of the IEEE Industrial Electronics Society, pp. 4472–4477 (2011)
9. Vogel-Heuser, B., Kegel, G., Bender, K., Wucherer, K.: Global Information Architecture for Industrial Automation. Automatisierungstechnische Praxis (atp) 51, 108–115 (2009)
10. The Industrial Internet of Things. Volume G5: Connectivity Framework (2017)
11. Theorin, A., Bengtsson, K., Provost, J., Lieder, M., Johnsson, C., Lundholm, T., Lennartson, B.: An Event-driven Manufacturing Information System Architecture for Industry 4.0. International Journal of Production Research 55, 1297–1311 (2016)
12. International Organization for Standardization (ISO): Information technology – Internet of Things Reference Architecture (IoT RA) (2016)
13. Delsing, J., Eliasson, J., Kyusakov, R., Colombo, A.W., Jammes, F., Nessaether, J., Karnouskos, S., Diedrich, C.: A Migration Approach Towards a SOA-based Next Generation Process

Control and Monitoring. In: IECON 2011 - 37th Annual Conference of the IEEE Industrial Electronics Society, pp. 4472–4477 (2011)

14. Castano, S., Antonellis, V. de: Global Viewing of Heterogeneous Data Sources. IEEE Trans. Knowl. Data Eng. 13, 277–297 (2001)
15. Industrial Automation Systems and Integration – Integration of Life-cycle Data for Process Plants Including Oil and Gas Production Facilities – Part 8: Implementation Methods for the Integration of Distributed Systems: Web Ontology Language (OWL) Implementation ISO 15926-8
16. European Committee for Electrotechnical Standardization (CENELEC): Application Integration at Electric Utilities – System Interfaces for Distribution Management – Part 1: Interface Architecture and General Recommendations (IEC 61968-1:2012) (2013)
17. European Committee for Electrotechnical Standardization (CENELEC): Energy Management System Application Program Interface (EMS-API) - Part 1: Guidelines and General Requirements (IEC 61970-1:2005) (2007)
18. Casado, R., Younas, M.: Emerging Trends and Technologies in Big Data Processing. Concurrency Computat.: Pract. Exper. 27, 2078–2091 (2015)
19. Marz, N., Warren, J.: Big Data. Principles and Best Practices of Scalable Real-time Data Systems. Manning, Shelter Island (2015)
20. Fan, W., Bifet, A.: Mining Big Data: Current Status, And Forecast to the Future. SIGKDD Explor. Newsl. 14, 1–5 (2012)
21. Begoli, E., Horey, J.: Design Principles for Effective Knowledge Discovery from Big Data. In: Babar, M.A. (ed.) Joint Working IEEE/IFIP Conference on Software Architecture (WICSA) and European Conference on Software Architecture (ECSA), 2012, pp. 215–218 (2012)
22. Institute of Automation and Information Systems, Technical University of Munich: SIDAP. Skalierbares Integrationskonzept zur Datenaggregation, -analyse, -aufbereitung von großen Datenmengen in der Prozessindustrie, http://www.sidap.de/
23. Lu, X., Li, Q., Qu, Z., Hui, P.: Privacy Information Security Classification Study in Internet of Things. In: 2014 International Conference on Identification, Information and Knowledge in the Internet of Things, IIKI 2014, pp. 162–165. IEEE (2014)
24. The Industrial Internet of Things. Volume G1: Reference Architecture (2017)
25. International Organization for Standardization (ISO): Information technology – Internet of Things Reference Architecture (IoT RA) (2016)
26. Chappell, D.: Enterprise Service Bus. O'Reilly Media, Inc. (2004)
27. Moser, T., Mordinyi, R., Winkler, D.: Extending Mechatronic Objects for Automation Systems Engineering in Heterogeneous Engineering Environments. In: Proceedings of 2012 IEEE 17th International Conference on Emerging Technologies & Factory Automation (ETFA 2012), pp. 1–8 (2012)
28. Klettner, C., Tauchnitz, T., Epple, U., Nothdurft, L., Diedrich, C., Schröder, T., Goßmann, D., Banerjee, S., Krauß, M., Latrou, C., et al.: Namur Open Architecture. Die Namur-Pyramide wird geöffnet für Industrie 4.0. Automatisierungstechnische Praxis (atp) 59, 20–37 (2017)
29. European Committee for Electrotechnical Standardization (CENELEC): Enterprise-control System Integration – Part 1: Models and Terminology (IEC 62264-1:2013) (2014)
30. Leitao, P., Barbosa, J., Pereira, A., Barata, J., Colombo, A.W.: Specification of the PERFoRM Architecture for the Seamless Production System Reconfiguration. In: Proceedings of the IECON2016 - 42nd Annual Conference of the IEEE Industrial Electronics Society, pp. 5729–5734 (2016)
31. Hufnagel, J., Vogel-Heuser, B.: Data Integration in Manufacturing Industry: Model-based Integration of Data Distributed From ERP to PLC. In: 2015 IEEE 13th International Conference on Industrial Informatics (INDIN), pp. 275–281 (2015)

32. Wang, Z., Dai, W., Wang, F., Deng, H., Wei, S., Zhang, X., Liang, B.: Kafka and Its Using in High-throughput and Reliable Message Distribution. In: 2015 8th International Conference on Intelligent Networks and Intelligent Systems (ICINIS), pp. 117–120 (2015)
33. Karnouskos, S., Bangemann, T., Diedrich, C.: Integration of Legacy Devices in the Future SOA-based Factory. IFAC Proceedings Volumes 42, 2113–2118 (2009)
34. Derhamy, H., Eliasson, J., Delsing, J.: IoT Interoperability - On-demand and Low Latency Transparent Multi-protocol Translator. IEEE Internet Things J., 1 (2017)
35. Vogel-Heuser, B., Legat, C., Folmer, J., Feldmann, S.: Researching Evolution in Industrial Plant Automation. Scenarios and Documentation of the Pick and Place Unit (2014)
36. International Organization for Standardization (ISO): Information Technology - Message Queuing Telemetry Transport (MQTT) v3.1.1 (2016)
37. OPC Foundation, https://github.com/OPCFoundation/UA-.NETStandardLibrary
38. Pivotal Software Inc.: RabbitMQ, https://www.rabbitmq.com/
39. Pivotal Software Inc.: Shovel plugin, https://www.rabbitmq.com/shovel.html

Social Science Contributions to Engineering Projects: Looking Beyond Explicit Knowledge Through the Lenses of Social Theory

Peter Müller[0000-0001-6603-3348] and Jan-Hendrik Passoth[0000-0003-1080-5382]

Technical University Munich, Arcisstraße 21, Munich 80333, Germany
{pet.mueller, jan.passoth}@tum.de

Abstract. With this paper, we will illustrate the synergetic potential of interdisciplinary research by demonstrating how socio-scientific perspectives can serve engineering purposes and contribute to engineering assignments. This especially concerns eliciting knowledge models and ergonomically optimizing technological design. For this purpose, we report on our findings on socio-technical arrangements within smart factory research initiatives that are part of the IMPROVE project. We focus our findings on the systemic interplay between the formally modelled plant, its actual physical state and the social environment. We also look at how operators, as parts of the plant's environment, adapt themselves and thereby develop their own particular work culture. We then integrate these findings by reconstructing this operator work culture as a specific way of performing, accounting for and addressing particular issues. We enhance these findings using the lenses of recent concepts developed in the field of social theory, namely the praxeological understanding of tacit knowledge, systems theory differentiations and an actor-network-theory understanding of human-machine agency. Applying these concepts from social theory, we revisit our empirical findings and integrate them to provide context-sensitive, socio-scientifically informed suggestions for engineering research on knowledge models concerning HMI design.

Keywords: Socio-technical Arrangements, smart industry, HMI design.

1 Introduction

Imagine you are in a factory and hear someone (or something) say: "Although this seems to be a batch problem, you should rather consider checking the rolls". With the contemporary smartification of industries, you cannot know whether you are hearing a human operator or a decision support system (DSS) speaking. The machines and devices in smart factories are no mere high-tech devices; they connect, cooperate and synergize with human actors. They are no simple tools of monitoring and control but a nexus of human-computer interaction (HCI), enhancing the capabilities and capacities of both human operators and the industrial plant complex. This development raises new concerns, issues and objectives of human machine interface (HMI) design: How to organize changing knowledge models? How to integrate informal intervention suggestions and formal automation? How to introduce new tools like smart glasses, tablets, smart watches? And, not

O. Niggemann and P. Schüller (Eds.), *IMPROVE – Innovative Modelling Approaches for Production Systems to Raise Validatable Efficiency*, Technologien für die intelligente Automation 8, https://doi.org/10.1007/978-3-662-57805-6_2

least, how to alter rather passive tools of knowledge assessment like dashboards and operating panels so that they include a close, pro-active entanglement of these features, rendering the smart factory, in terms of HCI, sensorimotor and self-reflexive?

However, these are well-known issues of HMI engineering, especially concerning smart, recursively learning software and human machine *cooperation*. Topics such as responsibility management, the avoidance of human errors and overcoming bounded rationality have been intensively discussed by HMI engineers and HCI scholars. [1] [2] With the critical momentum created by HMI software's ability to learn from its users' behavior, human-centered engineering is required and must be adjusted to the new assignments deriving from this significant technical shift: there are no longer any clear boundaries between passive technological tools and active users. While operators have developed their own particular ways of handling (their) HMIs, these interfaces themselves are beginning to be designed and act in a creative style, since they are the media technologies through which new patterns of knowledge and expertise can emerge and evolve.

Because of all this, new forms and sets of expertise for interdisciplinary research are being invoked. We claim that the social sciences can contribute a suitable perspective and methodology that goes beyond the canon of cognitive science, psychology and ergonomics that has been used so far to support and enhance HMI research. The social sciences are specialized in analyzing and understanding integral shifts of technology that affect not only work routines but the necessities of organization having to do with entrepreneurial, economic and personnel issues. Furthermore, the social sciences, like sociology and Science & Technology Studies (STS) in particular, are designed to tackle the kinds of complex entanglements that withstand technical approaches of proper explication, operationalization, formalization and trivialization. Patterns and meanings of complex work routines and conduct cannot be modeled by means of data mining for they cannot be operationalized accurately in numbers.

Being focused on the study of societal phenomena, the social sciences possess a specific repertoire of theories, concepts, methods and strategies for understanding the diverse and simultaneous dynamics of practices and situations that draw together knowledge, awareness, technology, incorporated routines, work cultures, organizational frameworks and business requirements. There are specialized disciplines for each of these aspects; however, when it comes to such complex assemblages with opaque effects and conditions, the social sciences can offer mediation and informed intuitions to supplement design processes and complement implementation scenarios.

2 Introducing our role(s) as social science researchers

2.1 What do social scientists do?

With the turn from HCI to human machine cooperation, it is necessary to draw stronger epistemological connections between the individual actors, experts and stakeholders involved in and affected by HMI development and assessment. Since the social sciences specialize in questioning the things that are taken for granted, they can thus help bridging the assumptive bounds that seem to analytically separate them. This conduct of questioning and looking out for the danger of implicit ontologies is accompanied by a sensitivity for latent, implicit structures, practices and knowledges. In the social sciences it is common to conceive 'knowledge' in plural terms because there are many different kinds, forms and constellations of knowledge(*s*). The peculiar perspectives, conduct, thinking and mindset of social science research have already been indicated. So far, we have claimed that this includes capacities which can contribute to engineering projects beyond technological expertise.

Socio-scientific perspectives have their own particular ways and strategies of research. Some readers will already have recognized that the social sciences not only have their own epistemological and methodological set(up), but also a specific style of thought and communication. Sociologists in particular tend to produce illustrative and passionate narrations instead of modest reports. This is not a (mere) result of the social sciences being so-called *soft sciences*, but rather reflects their genuine interest in issues of self-reflection and self-involvement. [3] Hence, we *tell* the *story* of our particular scientific role within and experiences with the IMPROVE project, for our experiences are a great part of – or at least complement to – our research material and practices. [4]

Social science research is empirical research; looking for the prerequisites of (inter)actions and practices, and that reconstructs seemingly non-social entities like technology as deeply entangled with social, cultural path dependencies and events. We therefore look for social artifacts (like documents, reports, communications) as well as for social processes like work practices, decision making, etc. Sociology stresses terms like *practices* to highlight that conscious and rational *modi operandi* are the exception, not the standard of how actors act. From this perspective the social and our everyday lives mostly consists of latent structures and behavioral routines that are usually reflected and *reasoned about only* retrospectively, not in advance. Thus, the social sciences have developed an elaborate terminology around latency and implicit patterns that shed some light on meanings embedded with social practices like work routines and HCI beyond the explicit dimensions of phenomena and data.

Another distinctive feature of empirical, socio-scientific research is that it regularly regards the social sciences regard qualitative research as sufficient for a certain subset of questions. Such research is characterized by small sample sizes, often have highly diverse data types (e.g. observations, documents, interviews) and an iterative, open (i.e. not fully formalized) methodological conduct. We organize, reflect and report our research process using the technique of *memoing*. *Memos* are a specific type of data: they are research notes that help to organize further research steps and which reconstruct the (subjective) research project to compensate this sociologically inevitable subjectivity. We process these data by *coding*, i.e. clustering different semantic commonalities. These commonalities are studied by means of comparison, organized by structural features of the smallest and greatest similarities/differences. As a result, we identify central concepts hidden in the data and codes which then structure our account of the phenomena in question. Social scientists handle the given lack of standardization by a theoretically informed and detailed description of how they reach their conclusion. [5][6][7]

2.2 What did *we* do as IMPROVE (social science) researchers?

As STS and sociology researchers, during our participation in the IMPROVE project, we investigated socio-technical arrangements in automatized, smart factories that applied

HMI and DSS. For this particular research, we collected three ethnographic meeting protocols, one complementary ethnographic protocol of the interview situation with our HMI partners, 29 project-internal documents (mental models, project plans, communication, HMI drafts, etc.), eight *memos*, and, both elicited by our HMI partners, three operator interviews, eight technician interviews.

Our designated role with IMPROVE was to investigate and reflect on the socio-technical arrangements involved with or stressed by the project's technological development or by smart factory environments. We were commissioned to derive a methodological toolkit from our research *experiences* which was intended to work as a *ready to use* methodology for further inquiries into socio-technical arrangements. Eventually, this is a methodological conundrum because socio-scientific methods often require practical training and can barely be standardized. Yet this was also an opportunity to develop strategies and precedents for *interdisciplinary diplomacy* [8]. For this purpose, we developed and set up a preliminary, socio-scientifically informed methodology that is formulated in a propaedeutic manner and takes into account discrepancies of disciplinary perspectives, demands and their matters of concern. Yet, this toolkit is rather a how-to-*become*-socio-scientific than a how-to-*do*-social-science. Furthermore, it has been our task to design a learning center for operators of smart factory plants. For this purpose, we made specific didactic suggestions regarding our own research on field-specific socio-technical arrangements and drafted tentative training scenarios and tools. Again, we brought in a *dash of sociology*, since we proposed not a positive, finalized learning center but a recursive *re*-learning center that would reflect the dynamic developments of both the usage of our toolkit and the matters of concern.

We we started our participation in the IMPROVE project with a review of project documents and the industrial engineering literature. We revisited the industrial sociology literature and papers on STS and the sociology of technology to deepen and broaden our understanding of the project's specific topics, issues and perspectives. We focused on requirement engineering, knowledge elicitation and modelling, as well as on HMI and DSS in smart factories and issues of human error and responsibility accounts. We thus focused on the role of tacit knowledge. By doing this, we discovered the fundamental discrepancy between engineering and sociological terminology mentioned in chapter one. However, we stressed the sociological or praxeological term of tacit knowledge, which contains *explicitly non-explicit* (and even non-explicable) knowledges (e.g. body movement, sensing practices or of latent structures and interrelations of communication). Hence, we concentrated our research efforts on the elicitation and exploitation of these technically neglected aspects of tacit, implicit practice knowledges.

We assisted and consulted the elicitation process and post-processed the empirical data (interview recordings, mental models, etc.). Inquiring [sic!] these empirical data, we identified central socio-technical arrangements. We also protocolled the project progress meetings in a socio-scientific manner to get our own understanding of engineering mindsets, assignments, communication and organization.

Methodologically, our research case produced two key concepts: "operators' work culture/particularization" and "explicity/implicity". These key concepts include codes like responsibility, routines, sensitivity and bounded rationality of operators, several industrial assignments, and the boundary between the formal, technological plant and the vast, vibrant plant environment (e.g. batch quality, machine quality, environmental parameters, organizational amendments, economic demands and limits). Thus, the physical plant itself is not identical to its formal, technological and scientific model, a technical discrepancy that is continuously bridged in terms of actualisation and contextualisation by human, social praxes. It is this realm beyond automatization which renders human operation necessary (aside from normative regulations for supervising high-risk processes). However, we wanted to avoid complacent sociological research that would exclusively follow the essential interests of our discipline. Therefore, we tried to work more towards applicable results and strategies of interdisciplinary diplomacy because these two assignments are interrelated: while interdisciplinary diplomacy is an applicable contribution, playing (with) the role of an applied science researcher supported our immersion in the disciplinary realms of our project partners. The results of this (re)search for a synergy of application orientation and mediation will be explained and illustrated in the subsequent chapters.

3 Empirical findings on socio-technical arrangements in HMI supported operating of smart factory plants

First of all, in social science, especially in qualitative research, theory is not always presented before empirical analysis. This is due to the fact that we distinguish between empirical and social theory. Social science, and quantitative research in particular, has empirical theories that have to be taken into account in advance in order to render deduction tests or explanations feasible. Qualitative social research, however, applies so-called *social theory*, which consists of non-empirical, theoretical considerations that give the social scientist perspectives, concepts, terms, suspicions or connections that can be used to organize the study and its findings according to the chosen social theory. This is a necessary part of socio-scientific research, since social phenomena are vastly complex and entangled in terms of synchronicity and overdetermination. Thus, classic deductive nomological models are insufficient, and simple, empirically grounded statements will fail to organize the socio-scientific knowledge and facts. Hence, social scientists apply similarly complex theoretical considerations in advance in order to identify significant data and draw plausible connections beyond everyday life plausibilities.

However, social theory is not the absolute *first step* of social science activities. It is rather integrated in an iterative research cycle, as also described by Grounded Theory. Grounded Theory is one of the first and greater disseminated proposals regarding how to do socio-scientific research in a way that generates theory from empirical data. [9] Such *grounded theories* are always, somehow, associated with social theory assumptions. Eventually, however, they must be integrated by means of greater social theories. Hence, social theory consists of both the presumed concepts that render socio-scientific observations or interpretations feasible and the conclusive integration of findings in a common terminology and conceptual perspective.

Nevertheless, we will now begin with a report on our empirical findings which will then be enhanced by social theory accounts. These social theories will elevate our qualitative findings on an analytical level of deeper understanding that can provide insights for technical design, planning and organization issues.

Concerning socio-technical arrangements, we encountered two significant features regarding human-operated industrial production. There is 1) a particular relation between a machine's formal system and its unstable environment. The formal machine system is the explicit, technological model that grounds any automation or overall technological design, e.g. the correlation between orifice width and film diameter, and the unstable environment is all those qualities which elude the idealized models due to their complexity (e.g. given climate variations and machine deterioration, market fluctuations). Hence, this complex environment is composed of physical and social entities and processes. In addition to this, operators established 2) a specific kind of work culture that derives from the intersection of historic, socio-economic subjectivity and their concrete task of *flexibly* operating these machines. We have already indicated that this is a necessity that results from the discrepancies between formalized technology and actual materiality.

Following our experience with and analysis of operator interviews, industrial plant operators have developed a work culture centered around particularity. This is no arbitrary quality, but a strategic outcome of the tension between the machine's formal system and its unstable environment. By work culture, we mean the self-descriptions, accountings and practical structuring of (re-)actions. We call it culture because it is a contingent result of adaptation between a specific (organizational) role within the given formal and practical limits of operators' work. While this work culture re-addresses itself as being particular to every individual, it organizes these individuals within their common situation through cultural presets of self-conception, praxis and perceiving. Thus, operator self-descriptions integrate an individualization of their role with the given requirements for common routines, semantics, etc. As a result of our tracing this work culture, we have learned that operators describe their own tasks as highly particular and often as incomprehensible from the exclusive perspective of formalization. Thus, the particularity produced by this work culture is a social reaction of personal and physical individuality confronted with bounded-rationality operating situations. This particularity results from plain psycho-physical, factual individuality and as a reaction to complex situations that cannot be reasoned beyond (individual) intuition.

However, there are also strategic reasons for operators to avoid describing their work as trivial so as to articulate their (technical) indispensability. Nevertheless, the resources of this self-presentation are no artifact of this strategy. Particularity within operating is rather a result of organizationally given frameworks (e.g. which adjustments are permitted to be done, which production assignments are set) and the actual, material complexity of the operated plant itself.

Thus, the operators who were studied produced a conglomeration of individualization figures, e.g. comparisons to driving a car (although it is a common practice, everyone has his own particular way of doing it) and explicit demarcations of the individuality and particularity of their work: Everybody has another way of handling this problem, everyone has his or her favorite way of solving such situations and issues, etc. Or, as they said: ask any operator about a specific problem and you will get *at least one* opinion, but probably more than one.

However, in contrast to these claims, operating work is, partially, more standardizable than depicted in these interviews, a fact which results in the further automatization of production plants. Automation is easily regarded as a threat of job loss by industrial workers. This conduct can result in rejection of automation innovations. Although this rejection is another issue, the practical results for operator practices and self-conceptions have to be taken into account. Nevertheless, there is a particularity which exists through the plant's realm of opaqueness, where different solutions can be applied, and thus different problems are identified. As a result, particularity is no simple self-display but part of the operators' practical self-specialization. However, there is a communality of operating which can be found on several levels (facility-wide, company-wide, etc.). This is indicated by the operators' shared terminology and conception of the plant. Beyond these facility- and individual-ordered cultures, other role-specific cultures and discrepancies can be found. These are marked by conflicts between technologists and operators in particular. There are, in general, latent conflicts and tensions which are expressed by the plain fact of how important it was, during our research, to guarantee that our elicitations were not and cannot be used for operator assessment. These conflicts represent a basic tension between educated, abstract and formal expertise on the one side, and practical, application-grounded expertise on the other. Intriguingly, during the experimental production scenarios, operating experience could not be successfully applied (as those scenarios were exceptional), nor could the formal models be completely confirmed. These conflicts can also be understood as tensions between those who design new technology and install and initialize plants, and those who maintain plants and keep them running. From our STS perspective, this is a particular socio-technical conflict of technologically shaped and undergirded social order: e.g. the responsibility of accounting for or reflecting organizational structures by means of technology access and the particular meanings that it has for its users. [10] [11] Thus, the tensions that appear when discussing a plant's reality happen between the two poles of formally understanding a machine complex and practically operating it in running scenarios.

These examples and observations are highly redundant in our given data. They applied to both case studies and they were repeated with each interviewed and observed operator. Even the one exceptional interviewee, an operator who was regarded as someone who shared the technological expertise of technologists and was familiar with both engineering expertise and practical efforts of operating, confirmed this particularity: whenever it comes to practically operating a plant, particularity is required or just results from the particularity of events.

Significantly, operators did not stress their particularity when describing their individual skill-sets. Instead, they particularized the description of their practices when it came to tacit knowledge requirements. They described their sensitivity towards the plant itself as being multi-sensitive: a feeling for the temperature and tension of the produced film, and even olfactory and acoustic signals were taken into account. Hence, their particularity cannot be reduced to mere expressions of an individualization discourse or a responsibility-handling strategy. It is first and foremost their practical way of operating plants beyond formality.

However, the operators we studied often told us about their indispensability, that not everything could be automated. This is, obviously, an expression of their own (economic) interest, highlighting their own importance within the industrial processes. Therefore, they often referred to the necessity of their work in terms of the limits of automatization. Human work cannot be completely replaced by software, they claim. From a neutral standpoint, automatization limits imply much more than the concerns uttered by operators. Those limits are less of a purely technological limitation than the economic limits of technological development (e.g. sensors) and legal or organizational limits (e.g. accountability requirements). Requirements that are not covered by contemporary software infrastructures include problems of accountability, responsible learning, etc. However, when it comes to tacit knowledge, the socio-technical arrangements of the plant itself and its human operators have their own path dependency and their own evolution: they are adapted to each other and their development cannot be understood without each other. Anything else would require a radical redesign.

This incorporated and situated, tacit knowledge contains many valuable capacities that could, in principal, be extracted from the field's practices. However, it cannot be transformed into formal, explicit knowledge and remains, thus far, a human peculiarity and privilege. While the use of machine and deep learning technologies enables machines to reach insights beyond human scales, they are still unable to smoothly integrate themselves into an ever-changing environment. This is a result of machines' strict mode of perception and due to limited real world training and test data. While their cognitive processing is impressive, they often ignore events and data that are out of their preset scope. Hence, they fail in terms of human perception since we are used to understanding and assessing everything from our very own perceptive scope. Nevertheless, this human perceptual worldliness must not neglect the greater sensory opportunities that machines present. However, neither engineers nor operators seem to regard human work in this way and are merely addressing an opaque complexity to render their industrial value and indispensability plausible.

4 Social Theory Plugins

We will briefly present three selected social theory perspectives in connection with our empirical findings. Since we want to avoid a simple eclecticism of theoretical possibilities, the selected concepts presented below are all commensurable with each other and will be reflectively interconnected in the final theoretical subsection.

Three different social theories were chosen concerning their specific theoretical and terminological accounts: systems theory (system/environment), practice theory (tacit knowledge), actor-network-theory (agency). We chose those conceptualizations for they each feature particular sensitivities towards the empirical field of our research.

First, we will introduce systems theory for its accounts and differentiation of system and environment can organize our research inquired socio-technical arrangements' complexity and take into account the difference between the assumed, formalized concept of an industrial plant and its actual physical existence.

Second, we will tackle tacit knowledge which already is a theoretical account of practice theory (or: praxeology). Tacit knowledge fills in the gap between formalized rules and models and the actual physical entities. It is a modus operandi which we encountered directly in our research field. With the lens of praxeology we will thus enhance our empirical findings.

Third and last, we will introduce actor-network-theory (ANT) and its theoretical account of "agency" which focusses on the relations that make action and behavior possible at all. Hence, we will use the concept of agency to shed (more) light on the difficult relationship of human-machine interaction and *co-operation* (in the rather descriptive, not normative sense). Such a concept of agency will allow us to integrate the other two social theories into one analytical canon for the concept of network driven agency touches both systemic complexity and the body, physicality orientated praxeology.

However, note that although we try to create a coherent theoretical setup for the analysis of our findings, it works rather heuristically and in terms of applied science or (socio-scientific) technique. From a genuine sociological point of view, however, this needs further theoretical elaboration and empirical substantiation. What we present here is, so far, rather an illustration and proof of concept.

4.1 A systems theory of (smart) factories

The social theory accounts we are first to introduce are 'systems' and 'environment' as they are defined by contemporary, sociological system theory. It is a conceptualization that focusses on process, time and distributed logics of cognition and perception. We will begin with a basic introduction of the central systems theory terminology: autopoiesis, system and environment. Also, we will highlight the cybernetic design of systems theory and how our empirical findings can be organized in autopoietic systems and their environments. This organization will then be used to structure our empirical findings and to identify its contained complexity requirements in particular.

The sociological *Systems Theory* that we stress here might remind engineers of cybernetics. There are, by design, many similarities and congruencies, since sociology's systems theory is, basically, sociologically applied cybernetics [12]. From a technological point of view, production plants are operatively closed systems and appear identical to their formal depiction. Any external event or stimulus that occurs within the scope of this formal setup of the plant will be reproduced as a part of the plant itself. Nevertheless, formal models can be applied, and smart factories are even capable of further processing those data points into their own behavioral model. When they detect an increase of heat, they can compensate in order to maintain the plant's functional homeostasis. However, what suffices for the formal system might be different from the needs of the actual physical plant. Required transformations run through the different systems with their frames of reference e.g. formal correctness and practical viability. This reference discrepancy might cause errors or confusions, but a forced breaching of the system bounds would lead to a malfunction: e.g. breaking down production for the sake of formal comprehension. This is an error typical of bureaucratic processes that tend to ignore non-bureaucratic semantics of logic.

Systems theory calls such selective reproduction processes *autopoiesis* [13] since the systems (e.g. plants) are marked and constituted by their own boundaries, which are constantly reproduced by the decision regarding whether something is part of the plant (or the system) or not. The very feature of sociological systems theory is that any event might be interpreted as an event of autopoiesis. This way, interconnections between entities become more complex, but environmental differences of meaning can be integrated and a system's stability can be explained by understanding its boundaries. Furthermore, unintentional effects can be traced back to their systemic source so that amendments do not lead to infinite sequences of (re)irritating the treated system. This radical systems theory conceptualization was actually developed in neurobiology and has been intensively and extensively redesigned in terms of social theory by Niklas Luhmann [14].

The technological design, development and maintenance of a plant are, however, second order cybernetics, which is a systemic observation that takes into account both the differentiated system and its acts of differentiation, including their residues. Adjusting and further developing components is a reflection on the functionality of the whole system, and operating is even more a matter of highlighting the external overview of the plant in terms of restoring its formal normality and applying (inter-)actions beyond the formal scope. However, with the development of formal models and their formalization grades, system boundaries are set. It is important to recognize that not only is a plant's common environment (socio-economic or material environments, i.e. markets, climate, batches, etc.) beyond the systemic scope of the plant, but so are parts of the plant that seem to belong to it itself, e.g. randomly deteriorating components. Furthermore, overdetermined correlations are part of the system's environment insofar as they cannot be rationalized and raise the risk of system's misunderstanding, i.e. being unable to continue its behavioral reproduction (e.g. if intervening is not optional but the chosen measure is only possibly correct).

Being aware of the systemic boundaries of a plant supports its conceptualization towards socio-technical assignments of organizing responsibilities, accountabilities and handling ergonomic assignments within different, complex environments. It signifies organizational and social demands as well as technological limitations that might be complemented through human work. Human operating work can thus be regarded as a complementary system that reproduces itself from everything that lies within the re-entry of the plant's system but falls out of its formal scope. 'Re-entry' describes how, for any system, it is necessary not only to differentiate between stimulus acceptance or non-acceptance (e.g. a reflection that might indicate an unclean film) but also between those events that concern this differentiation and those that do not (e.g. a bird flying over the facility).

With this, we encounter another feature of systems theory perspective: Even though cybernetics was, traditionally, a philosophy of control, modern systems theory no longer uses the term 'control' because of the system bounds that render immediate control impossible. Instead, regulation is defined as a specific, evolved arrangement of irritations. This requires interface systems that arrange different *irritations* (as action resources) in a way that renders operating and maintenance feasible. For smart factories, this also requires an HCI that can organize, foster and synergize knowledge from this irritation ecology.

However, this systems theory perspective can structure the different areas that are at stake or matter when it comes to technological and organizational or economic design. Regarding HMI and other technological design issues, HMIs are used as a materialized systems-interface. While this theoretical perspective will depict anything as inter-systemic irritations, HMI can take into account insights about a system's order by structuring its own knowledge- and decision-management in a way that is informed by identified system boundaries. This corresponds, for example, with required rule behavior, which addresses the physical plant as the plant's systems environment, etc. Note that the software/hardware and algorithms behind an HMI can only work as second order cybernetics if their system environment is both computably commensurable and can observe its first order systems under its own system code. In other words, second order cybernetics software requires explicit and expectable objects. However, second order cybernetics can easily be attained by integrating human actors. This can even be enhanced through HMI or DSS technology that is specifically designed not as a second order, but as a second cybernetic entity of order and organization, providing feedback to its surrounding systems (which are environmental towards the HMI itself) and shaping irritations in terms of translation by supplementing semantic interaction homeostasis, i.e. by identifying signals and interventions, or means and results. As a constructive act within itself, systems theory analysis reflects the engineers' own adaptation towards their use-cases and technologies, thus integrating constructive reflection into the theoretical framework of design: taking into account system functions and bounds without mistaking it for the differentiation of explicit and implicit data, knowledge, processes, structures, etc.

4.2 Tacit knowledge beyond explicity

The role of tacit knowledge for social science and also engineering purposes already has been highlighted above. However, this is a specific, elaborated social theory account, known as 'practice theory' which tries to resemble and understand social events and continuities as a result of incorporated knowledge and routines, bodies and their (physical) environment. Hence, we will begin with an illustration of how focusing on a corporal dimension works as a social science epistemology. After briefly describing the concept of tacit knowledge and its relation to routines, rules and explicit knowledge, we will elaborate on the tacit knowledge we found during our empirical research and which conclusions can be drawn by using the lenses of practice theory.

In contrast to engineering that is informed by ergonomics or cognitive science [15], sociologists make use of the more radical concept of implicit knowledge. While the engineering of knowledge modelling and knowledge-driven data mining understands implicit knowledge as proto-explicit knowledge hidden in someone's mind, the sociology of practice labels this hidden knowledge, which yet has to be elicited, as an explicit one. Engineers thus transform their not-yet-explicit knowledge into explicit knowledge using methods which only use knowledge-orientated protocols and are designed to literally make respondents *tell what they think* which highlights the boundaries between explicit and implicit knowledge. Such methods suffer from the fact that respondents usually cannot do their work or answer that way because they do not actually think consciously about what they do and are thus furtherly strained by thinking about what they might think while doing their work. What is easily mistaken for implicit knowledge of a practice is rather explicit knowledge that is reconstructed in hindsight (as most thoughts are anyway); the actual tacit knowledge of a practice is in its actual process and gets partially lost during its reconstructive explication.

Praxeological social theory tackle this issue by differentiating knowledge in terms of its form: can it be *transmitted* through textual media? If not, it is implicit or tacit knowledge. It is knowledge that is completely dissolved in unconscious bodily practices or in emergent behavior patterns (e.g. how we use our native language), or that is generally explicable but cannot be learned by using its explicit form (e.g. you cannot learn to ride a bike by merely observing it or reading a book about it) [16]. Since engineering is accustomed to processing explicit data, which often can be easily quantified or brought into clear logi(sti)c schemes, there is, quite frankly, a whole new world of knowledge out there.

Tacit knowledge can be elicited by qualitative observations and techniques of praxis immersion. While latent, emergent interaction patterns might be comprehended through mere long-term observations, some tacit knowledge might require a rather detailed, videographically supported method. Furthermore, some knowledge even to be elicited through participation in the practices which are studied. A large set of methodological suggestions and readjustments of perspective within sociology deals with eliciting, analyzing and interpreting tacit knowledge for empirical science purposes [17]. Therefore, HMI for operators in smart factories would benefit from such qualitative research. Socio-scientifically inspired qualitative research might lack objectivity and generalizability, but it can nevertheless make a significant difference in the quality of technology design.

However, tacit knowledge is ubiquitous and can be taken into account for all kinds of purposes. Any explicit knowledge requires whole sets of tacit knowledges, like grammar or an understanding of specific terminology. Much explicit knowledge is already contextualized and therefore is not regarded as involving tacit knowledge and can be taken for granted. When it comes to contextual practice, however, tacit knowledge is much more significant and visible (since it requires specific incorporated knowledge that might not be shared). This especially concerns rule-based practices like checking a machine's condition and being able to focus on particular, distinct features. Following this relation of tacit and explicit knowledge, acquiring tacit body knowledge can, indeed, be supported by explicit knowledge provisions. It might be required to present this explicit support in media that is not text-based, like videos or physical support (from a teacher or trainer). It is worth noting that learning and training tacit knowledge can be illustrated and investigated, again, by qualitative research. Therefore, for the commissioned learning center and training scenarios, we suggested a *relearning* center that would learn from its own practice-objective relations and their developments.

As we have discussed in the above section on our empirical findings, particularity is stressed and narratively constructed by operators whenever it comes to the application of tacit knowledge. With their particularity narrative, they can address those unspeakable things which defy plain explication. Since those particularized contents systematically fall out of the methodological scope of classic knowledge acquisition, the indicated content is structurally missed. It is important to note that not all tacit knowledge is indicated by communication or significant events or phenomena. Sometimes tacit knowledge has to be dug out of all the obviousness within a field. Actually, the indication of narrative self-description by particularization is kind of a field-specific situation, and in this specific case it might even lead to the most significant issues of practical knowledge. However, a lot more can be extracted from the field by observation methods or video analysis (if possible). The engineering differentiation of explicit and implicit knowledge is orientated towards explicit knowledge because engineering is focused on use cases. However, strong research on a field's tacit knowledge might not only contribute to positive technological developments but might also render observable complementary problem cases that indicate trans-intentional, perturbing or irritating disturbances that result from the use cases' positivity. For example, physical interfaces like smart glasses might interfere with safety requirements for wearing helmets, or be generally inappropriate if performing maintenance also requires dirty and rough work that might damage such equipment. We propose, in contrast, an additional *problem-case* focus that highlights not the positive technology but rather the frames of reference that situate and perturb socio-technical entities and connections.

In conclusion, HMI and DSS design can profit from this perspective and complementary research. Some content, like rules, can be put into the knowledge- and rule-organizing structure of the DSS, and some issues might be even capable of ordinal processing. And even beyond that, social sciences can highlight the expertise beyond explicit knowledge and help to think of an HMI and DSS that is able to synergize an operator's particular experience and further enhance their system regulation skills by providing feedback in the form of successful and handy solutions (e.g. applying a recommender system). Or, vice versa, an HMI and DSS would be able to warn of reported intuitive deceptions. Socio-scientifically informed research on tacit knowledge can also be applied for the purpose of automatization. Tacit knowledge not only takes additional, complementary data into account, it also reveals the complexity of such practices that seem to be simple routines, and it can thus contribute to modeling processes for automatization purposes. [18]

4.3 Conceptualizing human-machine agency

In this section we will introduce the theoretical account of 'agency' to better understand human-machine interaction. We will begin with an basic introduction of Actor Network Theory (ANT) and pragmatism (as it is applied by social theory). Since these social theories can appear to be counterintuitive, we will offer typical examples that highlight their analytical and interpretative value. We will then integrate our empirical findings into this theoretical account and furthermore conclude with a synthesis of the social theory accounts chosen for this study. [19]

This last sub-section reflects on the common conception of agency, of capabilities of being active or passive. HMI engineering has found that HMI software is not a mere tool which simply enhances its users' capabilities, but rather shapes its users' behavior and thus it can sometimes be risky if human actors do not correct machine errors [20]. Nevertheless, there is a theoretical perspective missing that can systematize these insights beyond mere empirical observations. There are several theories which focus their perspective on the issue of action networks, i.e. the interconnected entities which enable the whole network to act since no single entity would be able to act (for further illustration: imagine that, if you were the only thing that existed, there would be nothing to towards which you could possibly act). Although it is some decades old now, ANT [21] is still being discussed and has been further developed by many authors [22]. A similar, much older theoretical approach had been developed pragmatist philosophy [23].

As engineering usually focuses on use-cases, its tool-conception of technology affirms the common structure of activity and passivity. Insofar as human actors are regarded as detached entities who are just using technological tools, this was at least discipline-intrinsically self-sufficient. However, engineering is increasingly taking into account research on humans themselves. Humans become objects themselves when it comes to technological design and development. Cognitive science, ergonomics and psychology have restructured our conception of humans within engineering and thus require an appropriate theory that makes it possible to overcome the classical differentiation between tools and users, passivity and activity. And emerging engineering fields like human-centered software development prove that demand of a sociologically supplemented knowledge acquisition.

Agency is the result of those very difficult networks. These networks are no simple condition of action, but they are intersections of interplay and, in terms of systems theory, accountable as actions. Things also behave in correspondence with each other. There is no so-called "reflex arc", i.e. the hypothesis that a stimulus, the cognitive processing of a stimulus and a response were subsequently ordered and distinct phases of action of behavior. Rather it is a matter of being a sensorimotor nexus [24] that integrates the perceived environment into the organism. You can find therein the behavioral description of a cybernetic conception of human action once again. As described above, stability within complex practices exclusively derives from second-order cybernetics that can organize vibrant irritations. Since objective environments react depending on an organism's actions, this also goes the other way around. This means that meaning is defined as, first of all, a direct link between perception, conduct and behavior. If one encounters an emergency, one usually reacts way before one recognizes what is going on, and this is an indispensable feature, in terms of evolution.

This also touches on issues of social order. Hence, social order might be taken into account in HMI design. There are identifications of situations, of sets of stimuli and responses (which are situationally identical) that are taken for granted or as inevitable. As a result, deviance requires high cognitive effort (like acting weird on purpose). This is an important fact which was already stated by Durkheim [25] and concerns issues of responsibility, behavior control and raising consciousness.

However, a stimulus already requires sensorimotor sensitivity. Perceiving something, e.g. focusing the eyes, is no mere sensory task, but a task of motoric adjustment. The detection of a significant phenomenon (e.g. a threat) requires and involves specific, bodily conduct.

The important fact and lesson of this theoretical conceptualization is that those things we usually discuss as explicit knowledge (e.g. feelings, intentions, reasons, meanings) are not additional intermediators of stimuli and responses but are operationally identical. This perspective can be hard to grasp since this operational identity is separated in analytical reconstruction. This is what we address as *conscious thought*. We can, for example, flee spontaneously from a fire, but we will intellectually separate our motives, movements and means in retrospect. Thus, reflexive and practical thought is operationally separated, a fact that must be taken into account for scenarios that combine practical routines and mindful intervention.

However, there is an analytical meaning in this active-passive conception which stresses the role of motives and goals: active actors have goals while passive actors have none. Passive actors are therefore very predictable since they do not decide on or rearrange stimuli and responses (which are themselves also stimuli just as stimuli are at the same time also responses). Thereby, passive actors seem qualified as means. Nevertheless, within agency networks, goals and ends might be identified, but they do not actually contribute to a plain, objective description of events. Specifically, they might play a role but are superfluous regarding effective outcomes and can even lead in such detailed descriptions to infinite regressions of motives. A common example from ANT [26] is cars, which feature the security measure that they can only be driven if the seatbelts are fastened. In this scenario, the primary motive for buckling up before driving is the motivation to drive itself, while safety is its actual outcome. Such planned networks can nonetheless be manipulated, e.g. one could put the seatbelt behind one's back, but this would not really pay off and, furthermore, it indicates that there are other actors participating, like discourses of law, law enforcement and security. However, safety is no arbitrary outcome since some engineers have designed the car that way. Again, their motivation might nonetheless derive from legal prescriptions, economic benefits or specific engineering discourses. Legal measures might be mere reactions to faster driving cars. This development, again, can be identified as an effect of economic necessities and then be interpreted as a result of capitalism, which is, again, again (sic!) a result of necessities within a growing structure of work and goods distribution, and so on and so forth. Some technology might seem to have a specific goal and, in the end, appear to be the result of the motivation of some graduate students to get their PhDs.

In conclusion, this further illustrates the sociological concept of tacit knowledge, as it is beyond explicit reflection or consciousness, rendering any action and practice possible and feasible at all. Thus, our chosen theoretical accounts can be integrated. We understand tacit knowledge better when we follow the concept of distributed, synergetic agency and comprehend the adjustment of behavior and thought through cybernetic complexes and nexuses. For example, when we want to remember a thing, we just decide to do so (even though we are used to this image) but have to write a note in our calendars or knot a handkerchief since our whole self-control is not an immediate capacity but a strategical effort in arranging our own environment, rendering likely those (re)actions which we desire. Although quite different social theories, we can present the sensorimotor actor-networks as a certain synthesis of the social theories introduced above. Systems theory can be taken into account for the infrastructure and logistics of agency correspond with the inter-systemic irritations between any system and its environment: there is no general, central point of structure or meaning but a vibrant complex of practices and events that are less describable in terms of control but in terms of regulation, resilience and homeostasis.

Consequently, HMI is a huge part of human machine cooperation, and such an ontologically symmetrical perspective, as depicted in this section, can help to understand the agency that results from human machine interaction and can thus inspire HMI and DSS design considerations with ethnographic insights, contemplation and immersion [27]. For IMPROVE, we therefore delivered a qualitative methodological toolbox that grants access to structures of tacit knowledge and the difficult inter-*re*actions in HCI scenarios. Again, we cannot produce general knowledge for any (smart) industry scenario, but generic tools and strategies to handle each case in particular. The theoretical and disciplinary perspective we offered can, thus far, only provide general remarks on the design of HMI and DSS infrastructures since they have to reflect symmetrical agency: the DSS has to listen and to speak as well; like operators learn to use an HMI, an HMI has to learn to be used (or to use) operators. Since this also corresponds to the postulated particularity of operator work culture, these are initial, general instructions deriving from our research and reflection. A smart factory DSS must learn from situations and its operators and organize and synergize given content. A smart HMI and DSS must be coded counterintuitively, it has to produce vibrant dynamics that are by necessity reduced by the particular, practical relation between device and operator. Simply put: It has to start making 'stupid' requests because clever software is set up, but a smart industry setup has to follow the particular inter*re*actions of a particular individual in a particular environment. Standardization cannot synergize this particularity without erasing it. Hence, it has to be the consequence of operating, not its prerequisite. This way, by amplifying the required particularity of particular environments, systems and scenarios can, in the long run, become a viable way for further standardization if there are smart digital infrastructures that can assess, construct and disseminate collective operating knowledges and behavioral patterns.

HMI can restructure this set-up and smooth the transition from traditional to smart industry. HMI and smartification affect behaviour and perceptions, and both organize the interaction between operator and plant. If the HMI contains a DSS that offers not only knowledge and suggestions but learns from the dynamic interaction triangle between plant, operator and interface, smartification changes can turn a high-risk challenge to a form of incremental progress. This is an opportunity to take systems theory into account and to order cases, databases and other informative or instructive notes according to the identified systems and environments. Furthermore, it is possible to analytically separate tacit knowledge from complex, semi-implicit rule knowledge. That way, DSSs are able to reduce the over-determination of case structures: incidences can be differentiated and knowledge organized by the skills required, according to the identified systemic areas. If the physical plant environment interferes with the formal system, HCI is sufficient to resolve this loss of homeostasis. But operators can also enhance the HMI's knowledge organization (and can enhance high-tech interfaces fed with real-time data) and act as applicators and *logisticians* of tacit knowledge practices. With most HMI in smart factories, mechanical adjustments such as adjusting a V-belt, which required almost tacit body skills, were replaced by interfaces that represent machine movements and statuses through distinct, yet rather abstract, numbers. While this is a very good basis for further automatization, HMI should be designed in a way that makes it possible for operators to handle those issues that are beyond numbers or where it would be inappropriate to treat issues in a forceful, formalized way.

5 Summary and outlook

Summing up, we have introduced the specific perspectives, capacities and strategies of social science researchers. We have then presented our own involvement with the IMPROVE project. In the following sections, we illustrated our central empirical findings and created accounts and interpretations for them by means of social theory which consisted of three different theoretical accounts: systems, practice and agency. We concluded the last section with an integrative synthesis of these accounts, implying first instructive suggestions that could be distilled from our analysis.

We have depicted our findings concerning socio-technical arrangements within smartified industry. We began by focusing especially on the interplay between operators and HMI, and thus identified those environmental connections that are beyond a formal technological scope and which are located in social areas. Beyond basic environmental factors like climate, market, batch and social norms and organizational bureaucracy, the physical plant is also part of the plant's environment with regard to its technologically purified model.

We then sketched the application of three exemplary social theory perspectives to engineering objects and objectives and thereby demonstrated how they can contribute to engineering assignments and why they are appropriate candidates for interdisciplinary research projects. The concept of tacit knowledge, the practical body knowledge that lies beyond textual explicity and consciousness, can contribute new fields of knowledge to research concerning HMI design, knowledge models and the ergonomics of technological fields. Systems theory provides a particular perspective that makes it possible to systematically differentiate the factors of a plant's model that lie beyond proper formalization. It is even possible to use a DSS to categorize and synergistically integrate operating cases. The symmetrical, pragmatistic perspective of ANT is an appropriate way of thinking through HMI design since it can respect the interplay between operators and software and understand interaction and coproduced agency, which is, again, a way of integrating tacit knowledge in the technological design. Eventually, we drew connections between our observations and insights into operators' work culture.

With this paper, we have illustrated the capacity of the social sciences to contribute to engineering purposes, especially concerning HMI development. We have done this by applying a socio-scientific perspective in order to identify the described socio-technical arrangements and by cultivating a social theory-inspired perspective on the data and technological assignments. In doing so, we ignored thus far implicated, practical method issues. However, in order to successfully design an interdisciplinary project that combines engineering and the social sciences, and to be able to take tacit knowledge into account and make full use of social theory's potential, methodological readjustments and advancements are required. Since sociologists of practice are already focused on tacit knowledge, they have already developed a battery of methods like technography [28], specifically detailed interview interpretations or breaching experiments. Yet, those methods must be reinterpreted as particular tools for socio-scientifically informed engineering research. Also, issues of and strategies for field access have to be discussed and revisited. Hence, as an outlook, we want to suggest further methodological considerations and practical, tentative experiments in the form of interdisciplinary projects between the social sciences and engineering.

Acknowledgments

This project has received funding from the European Union's Horizon 2020 research and innovation programme under grant agreement No. 678867.

References

1. Hoc, J.-M.: From human - machine interaction to human – machine cooperation. In: Ergonomics (7) 43, pp. 833-843. Taylor and Francis, London (2010).
2. Hollnagel, E., Amalberti, R.: The Emperor's New Clothes. Or Whatever Happened To "Human Error". HESSD (2001).
3. Glaser, B. G.: Theoretical Sensitivity: Advances in the Methodology of Grounded Theory. Sociology Press, Mill Valley/California (1978).
4. Clifford, J., Marcus, G. E. (eds): Writing Culture: The Poetics and Politics of Ethnography. 25th Anniversary Edition. University of California Press, Berkeley, Calif. (2010).
5. Law, J.: Theories and methods in the sociology of science: an interpretative approach. In: Social Science Information 13, pp. 163–172, Sage, London (1974).
6. Strauss, A. I., Corbin, J. M.: Basic of Qualitative Research: Techniques and Procedures for Developing Grounded Theory. 2nd Edition. Sage Publicatoins, Thousand Oaks/California (1998).
7. Charmaz, K.: Constructing Grounded Theory: A Practical Guide Through Qualitative Analysis. Sage, London (2006).
8. Müller, P., Passoth, J.-H.: Engineering Collaborative Social Science Toolkits. STS Methods and Concepts as Devices for Interdisciplinary Diplomacy. In: Karafillidis, A., Weidner, R.: Developing Support Technologies. Springer International, Cham (2018, forthcoming) Interdisciplinary Diplomacy. Forthcoming.
9. Glaser, B. G., Strauss, A. L.: The Discovery of Grounded Theory. Strategies for Qualitative Research. Aldine, Chicago (1967).
10. Suchman, L., Blomberg, J. Orr, J., Trigg, R.: Reconstructing Technologies as Social Pracitce. In: American Behavioral Scientit (3) 43, pp. 392-408. Sage, London (1999).
11. Suchman, L., Trigg, R., Blomberg, J.: Working artefacts: ethnomethods of the prototype. In: British Journal of Sociology (2) 53, pp. 163-179, Routledge, London (2002).

12. Paetau: https://papiro.unizar.es/ojs/index.php/rc51-jos/article/view/790
13. Maturana, H., Francisco, V.: Autopoiesis and Cognition: the Realization of the Living. In: Cohen, R. S., Wartofsky, M. W. (eds) Boston Studies in the Philosophy of Science (42). D. Reidel, Dordrecht (1980).
14. Luhmann, N.: Essays on Self-Reference. Columbia University Press, New York (1990).
15. Bravo-Vásquez, D.-M., Sánchez-Segura, M.-I., Fuensanta, M.-D., Antonio, A.: Combining Software Engineering Elicitation Technique with the Knowledge Management Lifecycle. In: International Journal of Knowledge Society Research. 3 (1), pp. 1-13. IGI Global, Hershey (2012).
16. Turner, S.: The Social Theory of Pracitces: Tradition, Tacit Knowledge and Presuppositions. University of Chicago Press, Chicago (1994).
17. Adloff, F., Gerund, K., Kaldewey, D. (eds): Revealing Tacit Knowledge. Embodiment and Explication. Transcript, Bielefeld (2015).
18. Smith, V.: Ethnographies of Work and the Work of Ethnographers. In: Atkinson, P., et al. (eds) Handbook of Ethnography, pp. 220-233. Sage, London/Los Angeles (2007).
19. Passoth, J.-H.,Peuker, B.,Schillmeier, M. (eds): Agency without Actors? New Approaches to Collective Action. Routledge, London (2012).
20. Hoc, J.-M.: From human – machine interaction to human – machine cooperation. In: Ergonomics (7) 43, pp. 833-843. CRC Press, Boca Raton (2000)
21. Latour, B.: Reassembling the social: An introduction to actor-network-theory. Oxford University Press, Oxford (2005).
22. Law, J., Hassard, J. (eds.): Actor Netword Theory and After. Blackwell, Oxford (1999).
23. Mead, G. H.: The Definition of the Psychical. In: Decennial Publications of the University of Chicago (1) 3, pp. 77-112. Chicago University Press, Chicago (1903).
24. Dewey, J: The Reflex Arc Concept in Psychology. In: Psychological Review 3, 357-370. American Psychological Association: Washington (1896).
25. Durkheim, E.: The Rules of Sociological Method. The Free Press, New York/London (1982).
26. Latour, http://www.bruno-latour.fr/sites/default/files/P-67%20ACTOR-NETWORK.pdf.
27. Hughes, J., Randall, D., Shapiro, D.: From ethnographic record to System design. Some experiences in the field. Journal of CSCW (1) 3, pp. 123-141, Springer, New York (1997).
28. Schubert, C, Rammert, W.: Technografie. Zur Mikrosoziologie der Technik. Campus, Frankfurt a.M. (2006).

Enable learning of Hybrid Timed Automata in Absence of Discrete Events through Self-Organizing Maps

Alexander von Birgelen and Oliver Niggemann

Ostwestfalen-Lippe University of Applied Sciences, Institute Industrial IT,
Langenbruch 6, 32657 Lemgo, Germany
{alexander.birgelen,oliver.niggemann}@hs-owl.de
http://www.init-owl.de

Abstract. Model-based diagnosis is a commonly used approach to identify anomalies and root causes within cyber-physical production systems (CPPS) through the use of models, which are often times manually created by experts. However, manual modelling takes a lot of effort and is not suitable for today's fast-changing systems. Today, the large amount of sensor data provided by modern plants enables data-driven solutions where models are learned from the systems data, significantly reducing the manual modelling efforts. This enables tasks such as condition monitoring where anomalies are detected automatically, giving operators the chance to restore the plant to a working state before production losses occur. The choice of the model depends on a couple of factors, one of which is the type of the available signals. Modern CPPS are usually hybrid systems containing both binary and real-valued signals. Hybrid timed automata are one type of model which separate the systems behaviour into different modes through discrete events which are for example created from binary signals of the plant or through real-valued signal thresholds, defined by experts.
However, binary signals or expert knowledge to generate the much needed discrete events are not always available from the plant and automata can not be learned. The unsupervised, non-parametric approach presented and evaluated in this paper uses self-organizing maps and watershed transformations to allow the use of hybrid timed automata on data where learning of automata was not possible before. Furthermore, the results of the algorithm are tested on several data sets.

1 Introduction

Increasing product variety, product complexity and pressure for efficiency in a distributed and global production chain cause production systems to evolve rapidly: they become modular, can be parameterized and contain a growing set of sensors [1].

In order to enable European SMEs to face these challenges and to utilize new technical possibilities, the Horizon2020 project IMPROVE is aimed at developing user support functions in terms of self-diagnosis (i.e. condition monitoring,

O. Niggemann and P. Schüller (Eds.), *IMPROVE – Innovative Modelling Approaches for Production Systems to Raise Validatable Efficiency*, Technologien für die intelligente Automation 8, https://doi.org/10.1007/978-3-662-57805-6_3

predictive-maintenance) and self-optimization (e.g. energy optimization, output optimization).

Models can be constructed manually by an expert, but this is difficult, costly and time consuming in today's complex and evolving production plants [21]. Instead of only relying on human expertise and additional engineering steps formalizing the necessary knowledge, the tasks stated above will be taken on in a data-driven way [14] where models are learned automatically from the data. For anomaly detection, the live data from the plant is compared to the predictions of the learned model and deviations from the normal behaviour are classified as anomalous.

The type of model depends on a variety of conditions such as available signals, the task of the model and the overall nature of the system. Modern cyber-physical production systems (CPPS) usually are hybrid systems, meaning they comprise both binary and real-valued signals. In general, these dynamic systems are state based, for example the system's state is defined by its current and previous binary control signals, and the actions taken out are time dependent [13].

One approach to perform anomaly detection in such systems is to use hybrid timed automata [11] as a normal behaviour model which utilize discrete events to learn the system's normal behaviour. These events often cause so called mode or state changes in industrial plants, e.g. *conveyor is running* or *heater is on*, which result in different behaviour of the system in each mode, as shown in Figure 1. Changes in the binary control and sensor signal values of the system can be utilized as discrete events directly. Another way to get discrete events is to set thresholds on continuous signals but this requires additional expert knowledge.

Hybrid timed automata are well suited to learn the normal behaviour in terms of the modes/states, transitions and corresponding timings from data in an unsupervised manner by using the aforementioned binary control signals of a system. The real-valued signals are processed within the states using other types of models such as regression models or models which reduce the dimensionality of the real-valued data. An example for this is the nearest-neighbor principal component analysis (NNPCA) [5] which is used for the experiments in section 4. The NNPCA was chosen because it is used frequently in a variety of research projects in our institute.

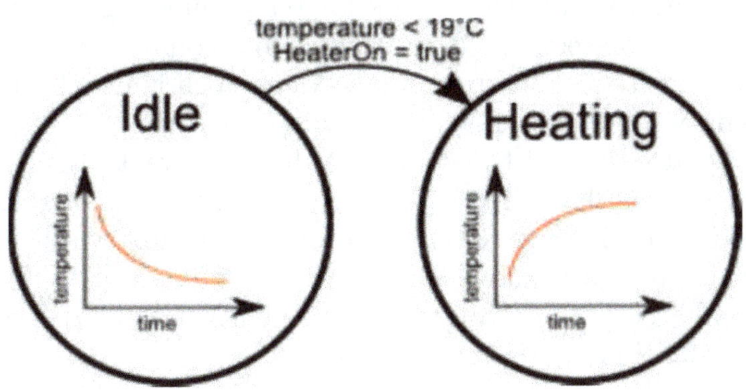

Fig. 1. An example of a hybrid automaton for a simple heating system with two states.

Unfortunately, expert knowledge about signal thresholds is almost never available and binary sensor signals are also not always available or not meaningful. This occurs for example when data is recorded using internal trace functionalities from drive controllers, which offer very high sampling rates but lack knowledge of variables from the programmable logic controller (PLC).

In this paper we present an unsupervised, non parametric approach to learn hybrid timed automata in absence of discrete events. The approach uses self-organizing maps (SOM) and watershed transformations to extract modes and generate discrete events in an unsupervised manner using only real-valued signals. The generated events can then used to capture the normal behaviour of the system by learning hybrid timed automata on data where they were not applicable before. The learned hybrid automaton is then used for anomaly detection in the real-valued signal values and in the time domain by analysing the transitions between the automaton's states.

The contents of this paper are structured as follows: section 2 introduces the existing modelling formalisms which are then combined in section 3 to generate discrete events from real-valued signals in an unsupervised manner. Experimental results from three different data sets, one artificial and two real world ones, are given in section 4. Finally, the paper is concluded in section 5.

2 Methodologies

2.1 Hybrid Timed Automata

Hybrid timed automata have proven to be a great tool to learn the normal behaviour of a system and detect deviations from it. Discrete events are required to learn an automaton. These events often cause mode changes in the system and the timing of these events is an important indicator for the health of the system. Hybrid timed automata are used to separate these modes, learn the transitions and timing between them and model the behaviour of the real-valued signals for each of the modes or states in the automaton.

An easy approach to obtain discrete events can be directly extracted from changes in the binary control and sensor signals of the system. It is also possible to obtain discrete events through thresholds for real-valued signals such as *temperature* $< 19^{\circ} C$ [7]. However, setting the thresholds and combinations of conditions for the continuous signals requires expert knowledge which is usually not available for real world automation systems. For unsupervised learning of these automata only binary control signals are used to obtain the discrete events, such as $HeaterOn = true$. Algorithms such as the online timed automaton learning algorithm (OTALA) [9] and its hybrid extension can work in an online, unsupervised manner, and do not require additional expert knowledge.

A hybrid automaton generated by the aforementioned algorithm can be defined as described in Definition 1.

Definition 1. *A hybrid timed, probabilistic automaton is a tuple* $A = (S, s_0, \Sigma, T, \delta, P, \theta)$, *where*

- *S is a finite set of states where $s \in S$.*
- *s_0 is the initial state which can be given by the systems state at the start of the training.*
- *Σ is the set of discrete events. Events $a \in \Sigma$ is linked to the transitions of the automaton.*
- *T is the set of transitions with $t \in T$ and $t = (s, a, s'))$, $s, s' \in S$ are source and destination state, $a \in \Sigma$ is the trigger event of the transition.*
- *The timing constraint $\delta : T \to I$ assigns a time interval to a transition $t \in T$, where I is a set of time intervals. The time here usually refers to the elapsed time since the last event occurred.*
- *P is a set of probabilities: for each transition $t \in T$ probability $p \in P$ is calculated.*
- *$\theta_{s \in S}$ describes a model for each state $s \in S$ which captures the behaviour of the real valued signals. Real valued signals are not captured by the discrete part of the automaton. These state models $\theta_{s \in S}$ are learned for each state of the automaton using other models such as linear regression, decision trees and others, such as the nearest neighbour principal component analysis (NNPCA) used in section 4 of this paper.*

The learned automaton can then be used to detect a variety of different classes of anomalies. This can for example be done using the anomaly detection algorithm (ANODA) [10] which can detect the following types of anomalies:

- **Unknown event / Wrong event sequence**: an event occurred which was not observed in the current state.
- **Timing error**: a transition occurred outside of the learned time bounds.
- **State remaining error**: when more time passed than for the latest event and the state is not a final state, then we have a state remaining error.
- **Probability error**: the probabilities of transitions for the new data are calculated and compared to the previously learned probabilities and an error is generated when deviations are too large.
- **Continuous error**: for each state, an additional anomaly detection for the real-valued signals can be performed using the internal state models.

2.2 Self-Organizing Map

The self-organizing map (SOM), also referred to as self-organizing feature map or kohonen network, is a neural network that can be associated with vector quantization, visualization and clustering but it can be used as an approach for non-linear, implicit dimensionality reduction [22]. The reduction is performed in a qualitative, implicit way. SOM's were chosen over classic clustering techniques due to their ability to find and visualize clusters in high dimensional data, which is often difficult with traditional clustering techniques [18]. A SOM consists of a collection of neurons which are connected in a topological arrangement which is usually a two dimensional rectangular or hexagonal grid. The input data is mapped to the neurons forming the SOM. Each neuron is essentially a weight vector of original dimensionality.

Definition 2. *the self-organizing map $SOM = (M, G, d)$ forms a topological mapping of an input space $O \subset \mathbb{R}^m$, $m \in \mathbf{N}$ and consist of*

- *a set of neurons M.*
- *each neuron $n \in M$ has a weight vector $\mathbf{w}_n \in \mathbb{R}^m, m \in \mathbb{N}$.*
- *G is usually a two-dimensional rectangular or hexagonal lattice in which the neurons $n \in M$ are arranged. Toroidal versions of these topologies are also common.*
- *$d(\mathbf{x}, \mathbf{y})$ is the distance measure to calculate the distance between two vectors $\boldsymbol{x}$ and $\boldsymbol{y}$ which can for example be weight vectors and/or vectors in the input space. Usually, the euclidean distance is used but other measures, such as the mahalanobis distance, can be used.*
- *an input sample $\mathbf{o}_i \in \mathbb{R}^m, i \in \mathbf{N}$ is mapped to the SOM through its best matching unit (BMU). The BMU is given by $bmu(\mathbf{o}_i) = \operatorname{argmin}_{k \in M} d(\mathbf{o}_i, \mathbf{w}_k)$*

One way to learn a SOM from data is a random batch training approach: The initial values of the neuron's weight vectors for the training can be randomly initialized or sampled from the training data. All samples from the training data are presented to the algorithm within one epoch. A best matching unit (BMU) is calculated for each input sample from the training data by finding the neuron which has the smallest distance to the sample. The BMU and all of its neighbouring neurons, assigned through the topology and neighbourhood radius, are shifted towards the input sample. Both the neighbourhood radius and strength of the shift decrease over time. The training stops after a chosen amount of epochs.

Each neuron of the SOM represents a part of the training data. Areas in the training input space with few examples are represented by few neurons of the SOM while dense areas in the input space are represented by a larger number of neurons.

The unified distance matrix (u-matrix) [19] allows a three-dimensional visual identification of clusters contained in the SOM. It calculates the sum of distances to neighbouring neurons according to the SOM's topology and visualizes clusters contained in the, usually high dimensional, training data. The neurons located on the borders of the non-toroidal SOM have fewer neighbours than the remaining neurons. Therefore, the summed distance is divided by the number of neighbours of the corresponding neuron to account for the different amounts of neighbouring neurons. The X and Y coordinates of the neurons represent the first two dimensions. The third dimension is given by the average distance to neighbouring neurons as in definition 3.

It can be visualized directly in 3D or in 2D using a color gradient as shown in Figure 2.

Definition 3. *for each neuron $n \in M$ and its associated weight vector $\mathbf{w}_n$, the u-matrix height is given by $U(n) = \frac{1}{|NN(n,G)|} \sum_{k \in NN(n,G)} d(\mathbf{w}_n, \mathbf{w}_k)$, where $NN(n, G)$ is the set of neighbouring neurons of n defined by grid G and $d(x, y)$ is the distance used in the SOM algorithm.*

The u-matrix representation illustrates why SOM's were chosen: SOM's tend to keep neurons with similar signal weights closely together. This results in a to-

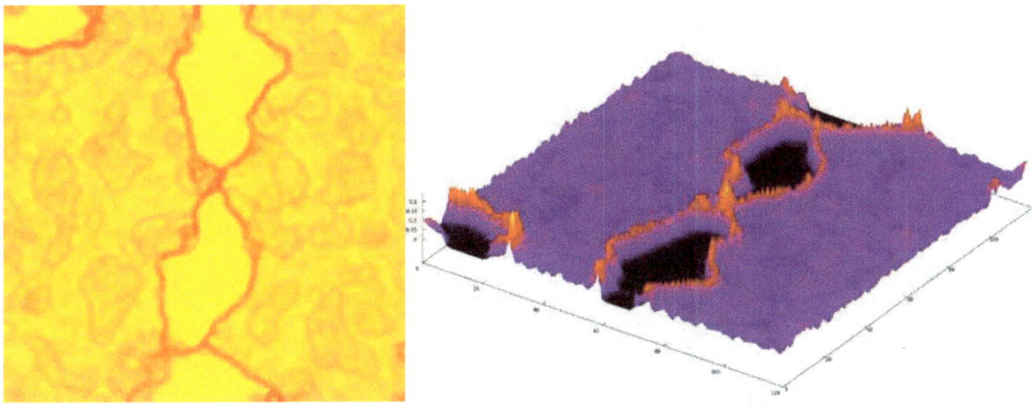

(a) Colored 2D (b) 3D

Fig. 2. Different u-matrix visualizations of a 120x120 SOM.

pographic landscape with valleys and ridges. Valleys represent clusters which are separated by the ridges.

Other works such as [8], [16] and [3] already performed anomaly detection on different processes by tracking the trajectory of the working point on the SOM: the observations are mapped to their corresponding BMU's as soon as they are recorded. Over time, the path or trajectory of the BMU can be observed and deviations from the known path indicate anomalous behaviour.

2.3 Watershed Transformation

The u-matrix representation allows visual identification of clusters in high dimensional data sets. This is easy to do for humans, but more difficult for a machine. Since the u-matrix representation can be plotted as an image, clustering algorithms from the image processing domain, such as the watershed transformation [12], can be used on the u-matrix representation of a SOM to identify the clusters in a mathematical way.

This works analogous to rain falling on top of the u-matrix. The water runs from higher regions to the lower regions and consequently flooding the basins. When the water level gets high enough so two basins merge, a ridge forms which separates them.

The watershed transformation dissects the u-matrix into different clusters, separated by the so-called watershed lines. Watershed lines separate the different basins and do not belong to any of the clusters. The basins can be interpreted as stationary process phases while the watershed lines represent transient process phases [6].

The implementation used here is the Vincent-Soille watershed algorithm which performs the watershed transformation in a non-recursive manner [20]. The sensitivity of the algorithm can be adjusted by setting a number of levels which in turn influences the number of final clusters found. Figure 3 shows examples using different levels on the same u-matrix.

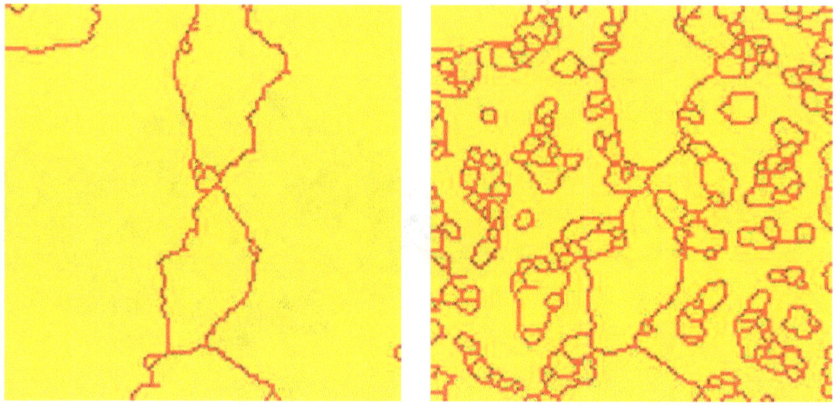

(a) 4 level, 16 cluster (b) 6 level, 197 cluster

Fig. 3. Watershed transformations of a u-matrix.

In the end, we receive a mapping for each neuron of the map to its corresponding cluster (Definition 4).

Definition 4. *the watershed transformation maps each neuron* $n \in M$ *to a cluster* c*, with* C *being a set of clusters and* $c = [0, |C| - 1] \in \mathbb{N}$.

3 Learning hybrid timed automata without discrete events

In order to learn hybrid timed automata, without binary control signals and without expert knowledge about thresholds for real-valued signals, it is necessary to derive the discrete events using an alternative way. For complex systems, the events can also be related to combinations of different real-valued signal values. Here, SOMs are used as a preprocessing step to extract the different modes of the system in an unsupervised manner.

As mentioned before, the trajectory of the BMU can be tracked and recorded. An automaton can be used to learn and track the trajectory of the BMU's. This can result in a large number of states in the automaton and many of the states might not contain enough data to learn a model from the data within the state. In this paper, we use the modes extracted from the SOM's u-matrix watershed transformation to group the data from the different modes and ultimately reduce the number of states of the automaton.

This section describes the generation of discrete events using SOMs. Figure 4 shows the general steps of the presented approach. A SOM is learned from the input data and the transitions between emerging clusters or rather system modes are used to generate discrete events. This is essentially a preprocessing procedure which allows learning hybrid timed automata in case the input data does not contain discrete events. After events are generated an automaton can be learned and then used to detect anomalies within the real-valued signal values and also within the timing and probabilities of the events.

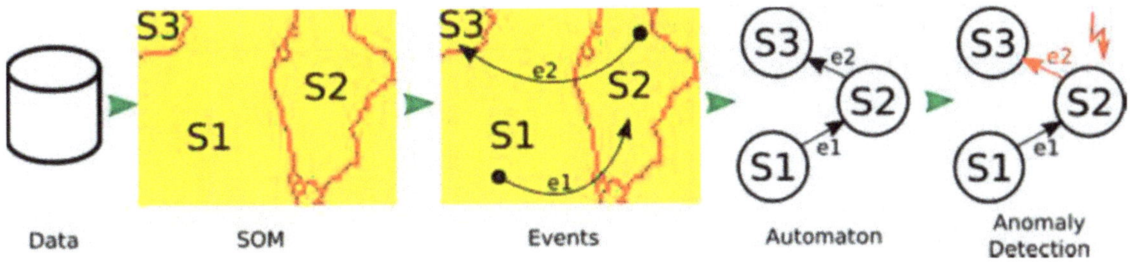

Fig. 4. General steps of the presented algorithm.

First, real-valued signals are recorded during normal operation of the plant to generate dataset O which consists of many obervations $o \in O$ and represents the input space of the model. The data can be seen as multivariate time series and each observation is equipped with a timestamp which is used when the automaton is learned. Then, the SOM-Discretization algorithm (Algorithm 1) generates discrete events for the data to allow learning of hybrid timed automata.

A SOM is trained on the recorded normal behaviour data (step 3). The size of the SOM is automatically calculated according to [17], where the the number of neurons $|M| \approx 5\sqrt{N}$ with N being the number of observations. The ratio of the side lengths is the ratio of the two largest eigenvalues of the data's covariance matrix. A normalization should generally be done before the SOM training but this depends on the input space and is therefore optional (step 2).

The SOM's u-matrix is calculated (step 4) and then clustered using the watershed transformation (step 5). The watershed levels are adjusted so the final cluster count is close to the shorter side length of the SOM. This approach seemed to work on all of the tested datasets (section 4 but this can probably be improved in future works. Each basin represents a stationary process phase and gets a unique number $i = [0, |C| - 1], i \in \mathbb{N}$ for identification. The watershed lines are transient process phase all receive a negative identification number to distinguish them from the stationary process phases. Observations mapped to transient process phases, the borders left after the watershed transformation, are assigned to the same cluster as the previous observation and therefore receive the same event vector as the previous observation (step 12). The event vector for cluster $c \in C$ in step 10 is created such that:

$$\mathbf{v}(c) = \left(v_1, v_2, ..., v_i, ..., v_{|C|-1} \middle| v_i = \begin{cases} 1 & \text{if } i{=}c \\ 0 & \text{else} \end{cases} \right)$$

The event vector is then concatenated with the original signal vector in step 14. The data now contain binary signals which are then interpreted as discrete events by hybrid automaton learning algorithms such as the hybrid OTALA used in this paper.

The anomaly detection can be performed either offline or online. Algorithm 2 shows the procedure for an online anomaly detection, where each observation o is tested as soon as it is received. When normalization was used for the training each new observation must also be normalized based on the normalization parameters

Algorithm 1 Generation of discrete events

```
1: procedure SOM-DISCRETIZATION(O)
2:     O ← NORMALISE(O)
3:     SOM ← TRAINSOM(O)
4:     U ← U-MATRIX(SOM)
5:     C ← WATERSHED(U)
6:     for all o ∈ O do
7:         bmu_o ← BMU(SOM, o)
8:         c_o ← C(bmu_o)
9:         if c_o ≥ 0 then
10:            v_o ← CREATEV(c_o)
11:        else
12:            v_o ← v_{o−1}
13:        end if
14:        p ← {v_o, o}
15:    end for
16: end procedure
```

calculated before (step 2). The best matching unit is calculated and linked to its corresponding cluster (steps 3, 4). When the observation falls onto a transient process phase it is assumed to be in the same cluster as the previous observation (step 8). When mapped to a stationary process phase, the discrete event signal vector is generated and appended to the observation in steps 6 and 10. The observation is then given to the ANODA algorithm to perform the anomaly detection using the learned automaton.

Algorithm 2 Preprocessing for online anomaly detection

```
1: procedure SOM-DISCRETIZATION-ONLINEAD(o)
2:     o ← NORMALISE(o)
3:     bmu_o ← BMU(SOM, o)
4:     c_o ← C(bmu)
5:     if c_o ≥ 0 then
6:         v_o ← CREATEV(c_o)
7:     else
8:         v_o ← v_{o−1}
9:     end if
10:    o ← {v_o, o}
11:    return o
12: end procedure
```

4 Experiments

This section presents experimental results of the presented approach using one artificial and two real world data sets.

4.1 Artificial test data

We created an artificial dataset through a simple PLC application. The PLC moves a virtual, linear SoftMotion [2] axis back and forth between two target positions. The drive uses trapezoid ramps with limited speed and accelerations. The drive related data from the software PLC are acquired through OPC-UA subscriptions in a 100ms publishing interval.

The data contains three real-valued signals: target position, actual velocity and actual speed. The training set contains 15 cycles, while the evaluation set contains five normal and five anomalous cycles. Maximum acceleration and deceleration are decreased during the anomalous cycles. An excerpt of the data is shown in Figure 5. The anomalies are labelled, to later evaluate the score for the anomaly detection. All of the machine learning methods here work unsupervised and have no knowledge about the labels.

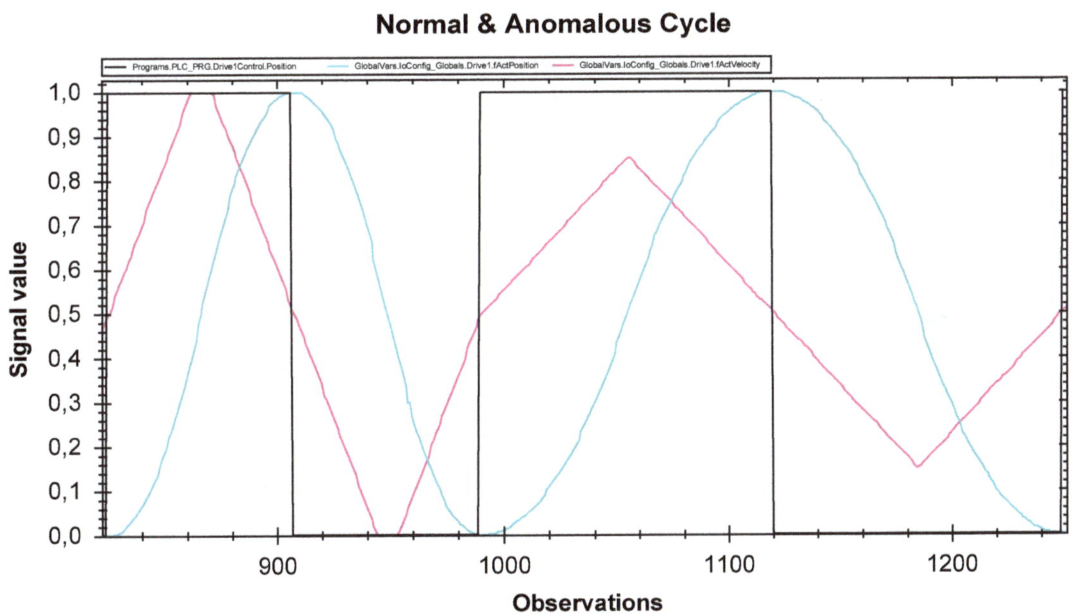

Fig. 5. One normal (827-991) and one anomalous (992-1251) cycle, scaled to range [0,1].

The nearest-neighbor PCA (NNPCA) [4] was first used on the full training data and then second as state-model for the hybrid automaton.

The data was reduced to two dimensions and then used for the anomaly detection. The training observations provide the reference to calculate the distance for each evaluation observation. When the distance to the reduced training space exceeds a certain threshold, the observation is considered anomalous. The threshold is calculated using a mexican-hat wavelet, converting the distance to an error probability [5].

Setting the threshold is not trivial and highly depends on the input data. A 100% threshold is good against false positives but also might be not sensitive enough to find the true positives. Lowering the threshold usually increases the true positive rate but at the risk of more false positives.

Figure 6 shows a plot of two-dimensional NNPCA learned from the normal behaviour used as training data. The anomalies from the evaluation data are unknown to the model and are shown here to visualize the separation between normal and anomalous behaviour. A threshold of 25% was selected, so every observations which gets a probability greater or equal to 25% of being anomalous is marked anomalous. Even with this low threshold, not a single anomaly is found by the NNPCA on the full dataset as the separation between normal and anomalous behaviour is not large enough to be found with the given threshold. Furthermore, the NNPCA model does not include time in an explicit way and therefore detection of anomalies in the time domain difficult or often not possible.

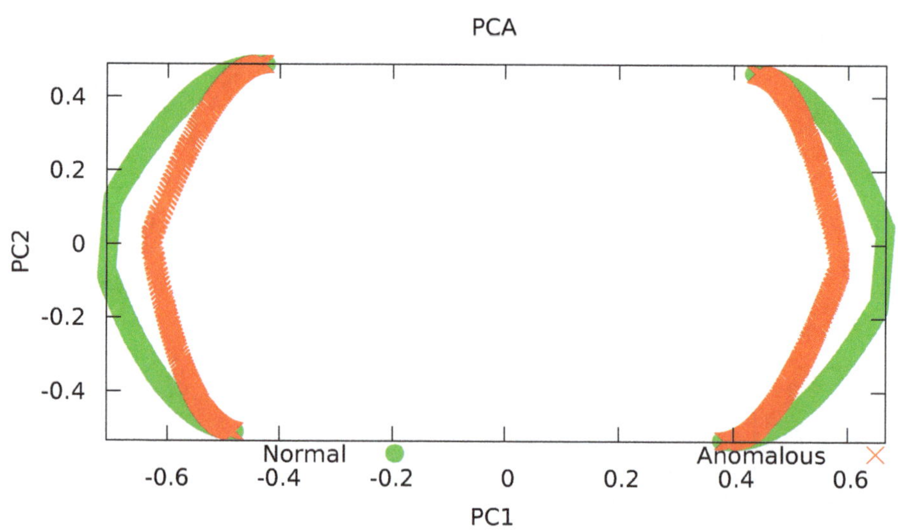

Fig. 6. Plot of the full dataset in the two-dimensional PCA space. Only normal behaviour is known to the learned model. Anomalies plotted to show the separation between normal and anomalous.

Now, algorithm 1 is used to generate discrete events so the data can be separated by a hybrid automaton. The automatic selection of parameters results in a SOM with 24x10 neurons and 11 clusters as shown in Figure 7.

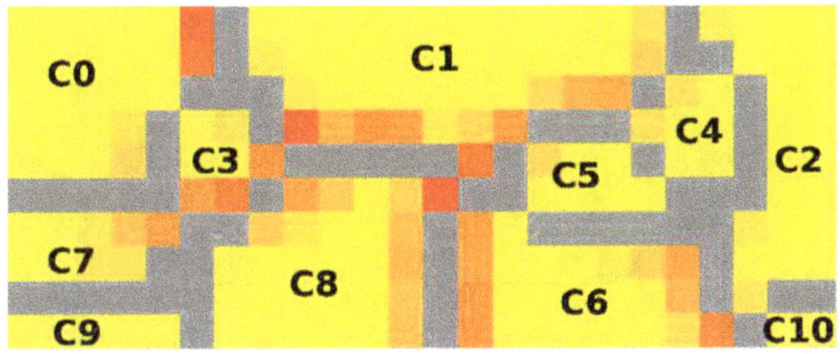

Fig. 7. Clusters on the SOM. Grey areas mark transient process phases.

With this, a hybrid automaton is trained using the hybrid OTALA algorithm and the NNPCA [5] as a state model with the exact same parameters as before. The resulting automaton is shown in Figure 8. Behind each state stands the state model from which two are shown in Figures 9a and 9b showing the separation of the normal and anomalous behaviour.

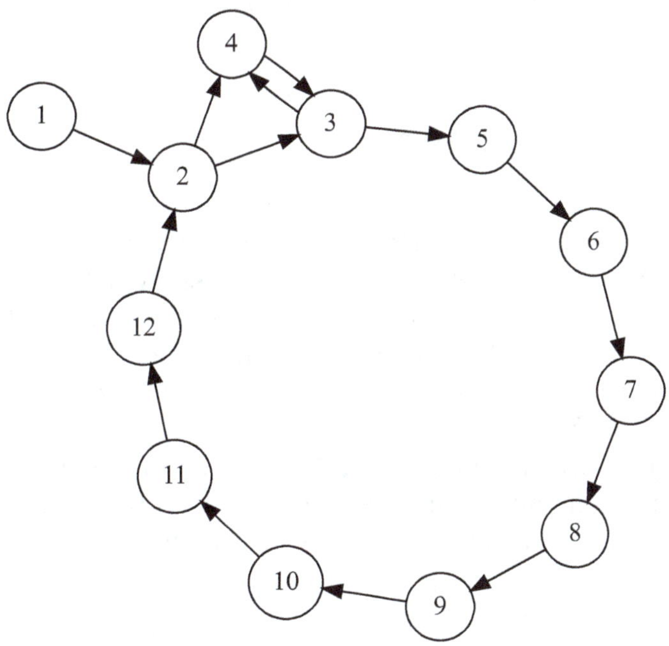

Fig. 8. The learned automaton with 12 states. State 1 is an initial state. Each cluster is mapped to one state.

The presented approach for the event generation in preparation for automaton learning calculates the necessary parameters from the training data as described in section 3. The settings for the NNPCA within the states were kept identical and the hybrid automaton achieves F1 measure of 97.53%, compared to the 0% F1 achieved by the NNPCA on the full dataset.

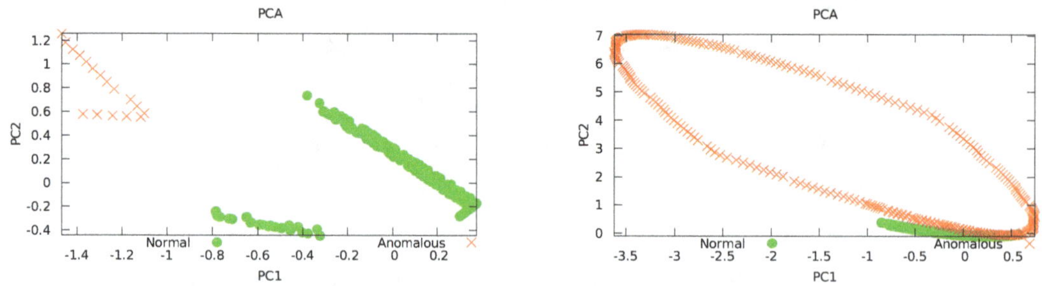

(a) State 3 (b) State 7

Fig. 9. Two state models from the automaton in Figure 8 showing the normal and mapped anomalous behaviour.

Table 1 shows some more details about the scores. The automaton offers another advantage: deviations in the timing and sequence of the system's behaviour can now be detected, which is not possible with the NNPCA alone.

4.2 High Rack Storage System

The High Rack Storage System or HRSS is a demonstrator from the SmartFactory-OWL which transports wares between its different shelves. The data from the system's drives includes position, power and voltage signals to a total of 17 real-valued signals, after removal of signals with zero-variance. The training data contains 107 cycles of the same transport operation. Evaluation data contains 113 cycles and was generated by modifying the training cycles in different ways such as increasing or decreasing one or multiple signals by different amounts (e.g. 20%), shortening of cycles and inserting constant sequences. This artificially generated anomalies are labelled and are used to calculate the scores of the anomaly detection.

The NNPCA reduces the data to 15 dimensions, keeping a variance coverage of 99.85%. A 60% threshold for the anomaly detection, successfully identifies 35 of the 7365 labelled anomalies resulting in an atrocious F1 score of 1.32%. The event generation creates a 34x22 SOM resulting in 23 clusters which are then captured by the automaton (Figure 10).

The automaton with the 15-dimensional NNPCA inside the states identifies 1794 true positives and reaches an F1 measure of 40.08% which is a drastic increase to the full-dataset NNPCA. It is to be noted that the anomalies were generated without respect to the overall dataset, so a 20% increase of a signal might not be significantly different from the overall training data and therefore it is possible that some of the anomalies are not significant enough compared to the whole dataset and might not be detectable.

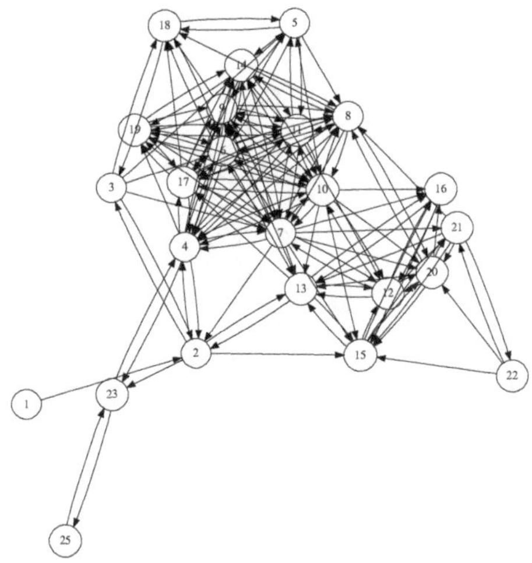

Fig. 10. The learned automaton for the storage system.

4.3 Film-Spool Unwinder

The third dataset presented in this paper was recorded from the cutting unit of a Vega shrink-wrap packer by OCME [15]. The machine groups loose bottles or cans into set package sizes, wraps them in plastic film and then heat-shrinks the plastic film to combine them into a package.

The drive controllers within the film cutting unit recorded chunks of data which each contain 2048 observations using their built-in scope functionality at a 4ms resolution. Here, we only consider two signals from the film unwinder drive: actual speed and lag error. 40 chunks were used for training and another 32 chunks were used as evaluation data. The film spool depleted during the last cycles, so every observation in the last 12 cycles was labelled anomalous. A dimensionality reduction is not necessary for two dimensions, but the NNPCA was still used to keep the model consistent for the experiments presented here. The NNPCA on the full dataset detected anomalous with an F1-measure of 18.63%. The automatically calculated size of the SOM is 64x22 neurons resulting in the automaton shown in Figure 11. The NNPCA on the full data and a selection of the 24 state models is shown in Figure 12.

The absolute scores for this dataset have to be taken with a grain of salt because the anomalies are not labelled observation-perfect but a generous range of observations was marked as anomalous. The F1 score for the anomaly detection with the automaton reaches 60.00%, resulting in an increase of roughly 41% (Table 1) compared to the NNPCA on the full dataset.

5 Conclusion

This paper presented an unsupervised, non-parametric approach which allows application of hybrid timed automata on data which do not allow their use due to a lack of knowledge about discrete events which are needed to learn and use hybrid

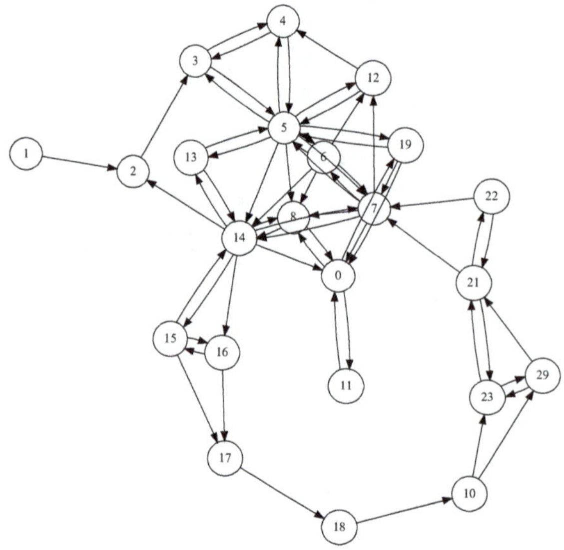

Fig. 11. Unwinder Automaton

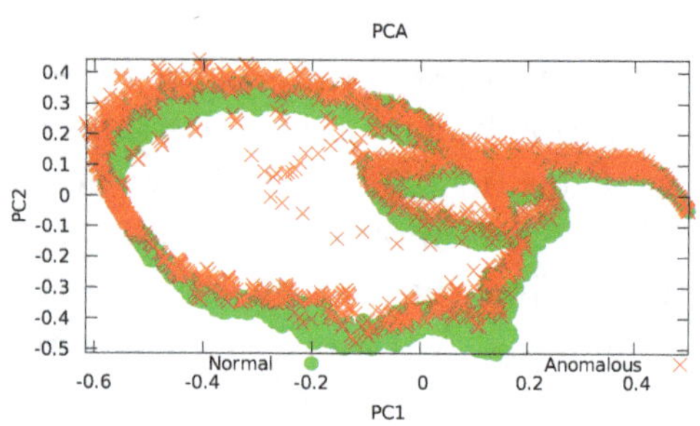

(a) PCA space of full data set

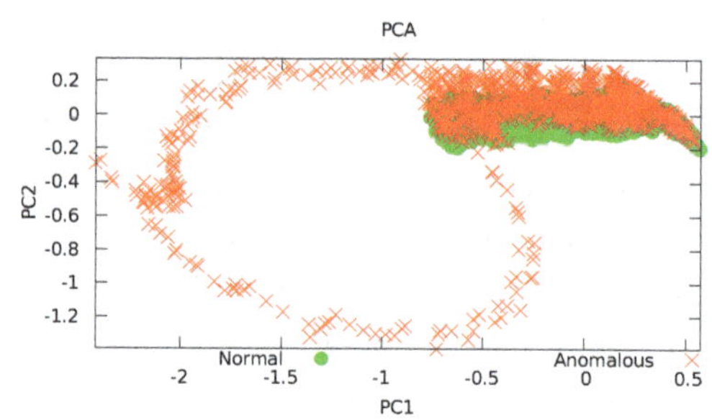

(b) State 3

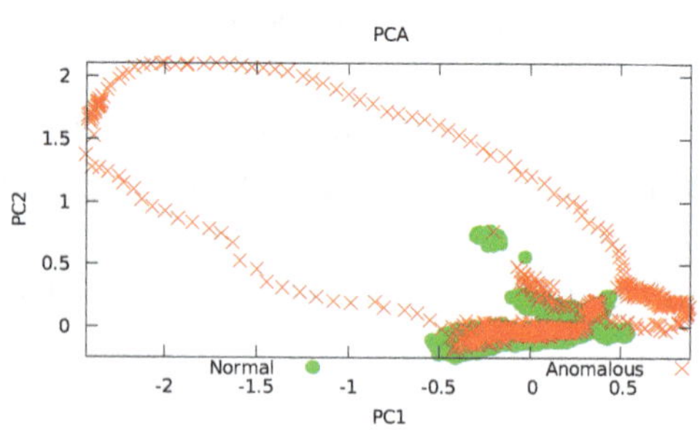

(c) State 7

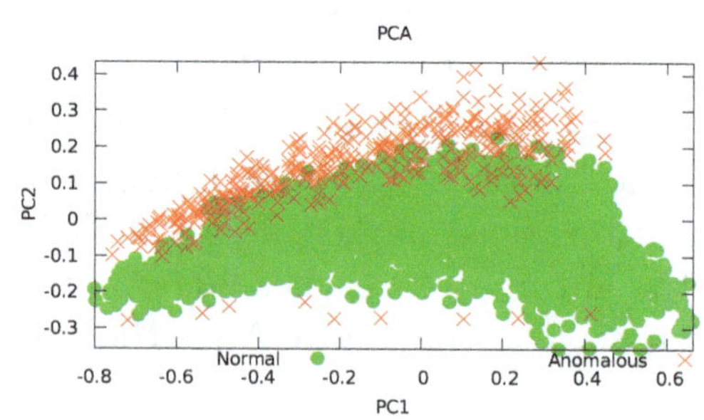

(d) State 18

Fig. 12. PCA of the unwinder data (12a) and examples from the automaton states (12b, 12c, 12d).

Table 1. Scores for different datasets and methods. Explanations can be found in the corresponding sections.

Dataset	Method	TP	TN	FP	FN	ACC	F1
4.1: Artificial	NNPCA	0	991	0	1692	36.94%	0%
	Automaton, NNPCA only	1568	990	1	124	95.34%	96.17%
	Automaton, all errors	1621	980	11	71	96.94%	97.53%
4.2: HRSS	NNPCA	35	15130	0	5239	74.32%	1.32%
	Automaton, NNPCA only	1636	13503	1627	3638	74.20%	38.33%
	Automaton, all errors	1794	13247	1883	3480	73.72%	40.08%
4.3: Unwinder	NNPCA	2528	40926	34	22048	66.31%	18.63%
	Automaton, NNPCA only	11340	37931	3029	13236	75.18%	58.24%
	Automaton, all errors	11864	37851	3109	12712	75.86%	60.00%

timed automata. Hybrid automata are used for anomaly detection in the real-valued signal values as well as in the time domain by analysing the transitions between the states of the automaton.

The presented approach derives the different stationary and transient process phases based on a system's real-valued signals through the use of self-organizing maps and watershed transformations. Also, all necessary parameters for the SOM and watershed transformation are automatically estimated from data.

The discrete events generated from the transitions between the extracted process phases are used to learn a hybrid timed automaton which in turn is used for anomaly detection. The presented algorithms work offline during the learning phase and can later be used online with live data from the plant. Further, the presented algorithms are not linked to a specific automaton learning algorithm and can be used as a preprocessing step prior to automaton learning.

The experiments in section 4 show results from three different datasets, two of which come from real-world machines. The anomaly detection for real-valued signals through a model within the states of the learned hybrid automaton increases the performance of the anomaly detection in comparison to the same model working with the same parameters on the full dataset. For all tested datasets, the separation of modes through the automaton improves the scores for the anomaly detection of the real-valued signals, without any additional tuning of parameters for the anomaly detection. Furthermore, models such as the aforementioned NNPCA or SOM's do not model time in an explicit manner. The hybrid automaton adds the explicit modelling of time for transitions between a system's modes and consequently enables the detection of anomalies which were not detectable before.

Acknowledgments. This project has received funding from the European Union's Horizon 2020 research and innovation programme under grant agreement No. 678867.

References

1. Factories of the Future: MultiAnnual Roadmap for the contractual PPP under HORIZON 2020. European Union, Luxembourg (2013)
2. 3S-Smart Software Solutions GmbH: Codesys softmotion: Integrierte bewegungssteuerung in einem iec 61131-3 programmiersystem (2017), https://de.codesys.com/produkte/codesys-motion-cnc/softmotion.html
3. Alhoniemi, E., Hollmn, J., Simula, O., Vesanto, J.: Process monitoring and modeling using the self-organizing map. Integrated Computer Aided Engineering 6, 3–14 (1999), http://citeseerx.ist.psu.edu/viewdoc/summary?doi=10.1.1.33.573
4. Eickmeyer, J., Krueger, T., Frischkorn, A., Hoppe, T., Li, P., Pethig, F., Schriegel, S., Niggemann, O.: Intelligente zustandsberwachung von windenergieanlagen als cloud-service. In: Automation 2015. Baden-Baden (Jun 2015)
5. Eickmeyer, J., Li, P., Givehchi, O., Pethig, F., Niggemann, O.: Data driven modeling for system-level condition monitoring on wind power plants. In: Proceedings of the 26th International Workshop on Principles of Diagnosis (DX-2015) co-located with 9th IFAC Symposium on Fault Detection, Supervision and Safety for Technical Processes (Safeprocess 2015), Paris, France, August 31 - September 3, 2015. pp. 43–50 (2015)
6. Frey, C.: Monitoring of complex industrial processes based on self-organizing maps and watershed transformations. In: Industrial Technology (ICIT), 2012 IEEE International Conference on (Mar 2012)
7. Henzinger, T.A.: The theory of hybrid automata. In: Proceedings 11th Annual IEEE Symposium on Logic in Computer Science. pp. 278–292 (Jul 1996)
8. Liukkonen, M., Hiltunen, Y., Laakso, I.: Advanced monitoring and diagnosis of industrial processes. In: 2013 8th EUROSIM Congress on Modelling and Simulation. pp. 112–117 (Sept 2013)
9. Maier, A.: Online passive learning of timed automata for cyber-physical production systems. In: 12th IEEE International Conference on Industrial Informatics (INDIN). pp. 60–66 (July 2014)
10. Maier, A.: Identification of timed behavior models for diagnosis in production systems. Ph.D. thesis, Paderborn, Univ. (2015)
11. Maier, A., Niggemann, O.: On the learning of timing behavior for anomaly detection in cyber-physical production systems. In: International Workshop on the Principles of Diagnosis (DX). Paris, France (Aug 2015)
12. Meyer, F.: Topographic distance and watershed lines. Signal Processing 38(1), 113–125 (jul 1994), http://dx.doi.org/10.1016/0165-1684(94)90060-4
13. Niggemann, O., Maier, A., Just, R., Jäger, M.: Anomaly detection in production plants using timed automata : Automated learning of models from observations. In: Proceedings of the 8th International Conference on Informatics in Control, Automation and Robotics. pp. 363 – 369. No. 1 (2013)
14. Niggemann, O., Maier, A., Vodencarevic, A., Jantscher, B.: Fighting the modeling bottleneck - learning models for production plants. Workshop "Modellbasierte Entwicklung Eingebetteter Systeme" (MBEES) (7 2011)
15. OCME: Shrink-wrap packers vega (Mar 2017), http://www.ocme.com/en/our-solutions/secondary-packaging/vega
16. Simula, O., Kangas, J.: Process monitoring and visualisation using self-organizing maps (1995)
17. Tian, J., Azarian, M.H., Pecht, M.: Anomaly detection using self-organizing maps-based k-nearest neighbor algorithm. Second European Conference of the Prognostics and Health Management Society 2014 (2014)

18. Ultsch, A., Ltsch, J.: Machine-learned cluster identification in high-dimensional data. Journal of Biomedical Informatics 66, 95 – 104 (2017), http://www.sciencedirect.com/science/article/pii/S153204641630185X
19. Ultsch, A., Siemon, H.P.: Kohonen's self-organizing feature maps for exploratory data analysis. In: Proceedings of the International Neural Network Conference (INNC'90 (1990), http://www.uni-marburg.de/fb12/datenbionik/pdf/pubs/1990/UltschSiemon90
20. Vincent, L., Soille, P.: Watersheds in digital spaces: an efficient algorithm based on immersion simulations. IEEE Transactions on Pattern Analysis and Machine Intelligence 13(6), 583–598 (Jun 1991)
21. Vogel-Heuser, B., Diedrich, C., Fay, A., Jeschke, S., Kowalewski, S.and Wollschlaeger, M., Goehner, P.: Challenges for software engineering in automation. Journal of Software Engineering and Applications 7(5) (May 2014), http://dx.doi.org/10.4236/jsea.2014.75041
22. Yin, H.: The Self-Organizing Maps: Background, Theories, Extensions and Applications, pp. 715–762. Springer Berlin Heidelberg, Berlin, Heidelberg (2008)

Anomaly Detection and Localization for Cyber-Physical Production Systems with Self-Organizing Maps

Alexander von Birgelen and Oliver Niggemann

Ostwestfalen-Lippe University of Applied Sciences, Institute Industrial IT, Langenbruch 6, 32657 Lemgo, Germany
{alexander.birgelen,oliver.niggemann}@hs-owl.de
http://www.init-owl.de

Abstract. Modern Cyber-Physical Production Systems provide large amounts of data such as sensor and control signals or configuration parameters. The available data enables unsupervised, data-driven solutions for model-based anomaly detection and anomaly localization: models which represent the normal behavior of the system are learned from data. Then, live data from the system can be compared to the predictions of the model to detect anomalies and perform anomaly localization. In this paper we use self-organizing maps for the aforementioned tasks and evaluate the presented methods on real-world systems.

1 Introduction

Modern Cyber-Physical Production Systems (CPPS) evolve rapidly, become modular, can be parameterized and contain increasingly more sensors due to increasing product variety, product complexity and pressure for efficiency in a distributed and globalized production chain they become modular, can be parameterized and contain a growing set of sensors [1]. This also means it becomes more and more difficult to monitor the systems. Human operators often struggle to diagnose faults or anomalous behavior in the system in time, leading to system break down, unexpected downtime or degradation in product quality.

A common approach to detect the aforementioned scenarios is to construct models for a given system and compare the predictions of the model to the real system. Anomalous behavior is detected when the real system's behavior deviates from the model's predictions. Manual construction of system models by experts is usually time consuming, expensive and also difficult in today's evolving complex systems. Experts with the necessary knowledge are usually scarce and often times some of the necessary knowledge is not available at all. Modern CPPS often provide large amounts of data such as control signals, sensor signals and configuration parameters [10]. This allows the use of data-driven methods: models are learned from data and then used for various tasks such as anomaly detection and anomaly localization.

O. Niggemann and P. Schüller (Eds.), *IMPROVE – Innovative Modelling Approaches for Production Systems to Raise Validatable Efficiency*, Technologien für die intelligente Automation 8, https://doi.org/10.1007/978-3-662-57805-6_4

Live data from the system is compared to the predictions of the learned model. Deviations from the normal behavior are classified as anomalous. Once anomalies are found, the anomalous samples are presented to a reverse model to localize the anomalies. This provides a starting point for plant operators and experts to restore the system to normal working order, ideally before production losses occur.

In this paper we use self-organizing maps (SOM) to learn a systems normal behavior in an unsupervised manner. The learned SOM's are then used for both anomaly detection and anomaly localization. The contents of this paper are structured as follows: First, section 2 explains the general concept of self-organizing maps. Second, section 2.1 presents an approach to detect and localize anomalies within the signal domain of a system. Third, section 2.3 introduces an approach where timed automata are used to track the working point on top of the self-organizing map in order to detect anomalies in the time domain. Furthermore, the aforementioned approaches to anomaly detection and localization are applied and explained on the Institute Industrial IT's OPAK demonstrator [11] in section 3. Section 4 presents the conclusion and future points of research.

2 Self-Organizing Map

The self-organizing map (SOM) [5], also referred to as self-organizing feature map or Kohonen network, is a neural network that can be associated with vector quantization, visualization and clustering but can also be used as an approach for non-linear, implicit dimensionality reduction [17]. A SOM consists of a collection of neurons which are connected in a topological arrangement which is usually a two dimensional rectangular or hexagonal grid. The input data is mapped to the neurons forming the SOM. Each neuron is essentially a weight vector of original dimensionality but provides additional information such as its coordinates within the grid. All experiments in this paper use a two dimensional, non-toroidal rectangular lattice and the Euclidean distance measure as shown in Definition 1.

Definition 1. *The* $SOM = (M, G, d)$ *forms a topological mapping of an input space* $O \subset \mathbb{R}^m$*,* $m \in \mathbb{N}$ *and consist of*

- *a set of neurons* M*.*
- *each neuron* $n \in M$ *has a weight vector* $\mathbf{w}_n \in \mathbb{R}^m, m \in \mathbb{N}$*.*
- G *is a two-dimensional rectangular lattice in which the neurons* $n \in M$ *are arranged.*
- $d(\mathbf{x}, \mathbf{y})$ *is the distance measure to calculate the distance between two vectors* $\boldsymbol{x}$ *and* $\boldsymbol{y}$ *which can for example be weight vectors and/or vectors in the input space. The euclidean distance is used for all models in this paper.*
- *an input sample* $\mathbf{o}_i \in \mathbb{R}^m, i \in \mathbb{N}, m \in \mathbb{N}$ *is mapped to the SOM through its best matching unit (BMU). The BMU is given by* $bmu(\mathbf{o}_i) = \operatorname{argmin}_{n \in M} d(\mathbf{o}_i, \mathbf{w}_n)$

One way to learn a SOM from data is a random batch training approach: the initial values of the neuron's weight vectors for the training can be randomly initialized or sampled from the training data to provide a diverse starting point for the training process. Training takes place over a chosen amount of epochs. All samples from the training data are presented to the algorithm within one epoch. A best matching unit (BMU) is calculated for each input sample from the training data by finding the neuron which has the smallest distance to the sample. The BMU and all of its neighboring neurons, assigned through the topology and neighborhood radius, are shifted towards the input sample (Figure 1). Both the size of the neighborhood and strength of the shift decrease over time to help with convergence. In the end, each neuron of the SOM represents a part of the training data. Areas in the training input space with few examples are represented by few neurons of the SOM while dense areas in the input space are represented by a larger number of neurons. Usually, the number of neurons is chosen much smaller than the number of samples in the training data, effectively discretizing and reducing the training data to the most important samples.

Usually, the number of neurons is chosen much smaller than the number of samples in the training data, effectively discretizing and reducing the training data to the most important samples. The compact representation of the training data provided by the SOM is used in section 2.1 to detect anomalies by calculating the quantization error.

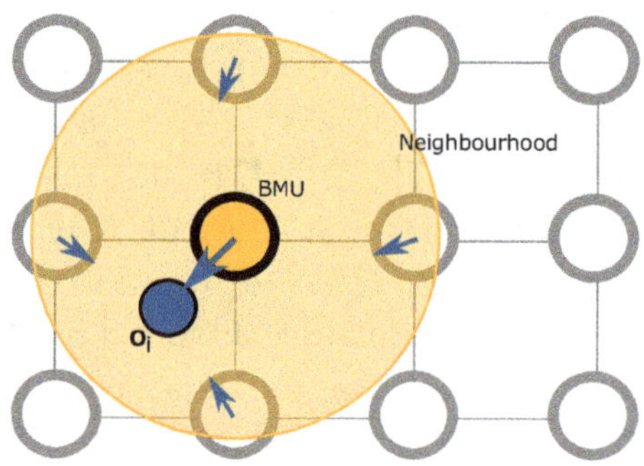

Fig. 1. Illustration of training procedure for a single sample.

The unified distance matrix (u-matrix) [14] of the SOM is great for visual identification of clusters in high-dimensional data. The distance to neighboring neurons according to the SOM's topology is computed and plotted as an image: the X and Y coordinates of the neurons represent the first two dimensions. The third dimension is given by the sum of distances to neighboring neurons as in definition 2. It calculates the sum of distances to neighbouring neurons according to the SOM's topology and visualizes clusters contained in the, usually high dimensional, training data. The neurons located on the borders of the non-toroidal SOM have fewer neighbours than the remaining neurons. Therefore, the summed distance is divided by the number of neighbours of the corresponding neuron to account for the different amounts of neighbouring neurons. A color gradient can be used instead of a third dimension as shown in Figure 3: valleys are represented by the color yellow, indicating a low distance between neighboring neurons. Ridges are represented by the color red, indicating a high distance between neighboring neurons. The u-matrix can be defined as follows in Definition 2.

Definition 2. *for each neuron $n \in M$ and its associated weight vector $\mathbf{w}_n$, the u-matrix height is given by $U(n) = \sum_{k \in NN(n,G)} d(\mathbf{w}_n, \mathbf{w}_k)$, where $NN(n,G)$ is the set of neighboring neurons of n defined by grid G and $d(x,y)$ is the distance used in the SOM algorithm.*

The u-matrix representation illustrates why SOM's were chosen: SOM's tend to keep neurons with similar signal weights closely together, which results in a topographic landscape with valleys, where weights of neighbors are similar, and ridges, where weights of neighbors are not similar. Valleys represent regions where the contained neurons weight vectors are very similar. These valleys are separated by ridges which mark transitions between the different feature spaces. In section 2.3 we further explore this matter to detect anomalies within the time domain.

2.1 Anomaly detection with quantization error

The SOM can be used to detect anomalies by calculating the quantization error: small errors below a threshold are considered normal, while errors above are considered anomalous. Quantization error based approaches were already used for tasks such as network monitoring [6] and anomaly detection in industrial processes [12][3][13]. These works however, did not perform an anomaly localization and only [13] used the quantization error as a measure for system degradation.

The quantization error (Definition 3) of each sample is calculated by mapping it to the SOM to get its BMU. The distance of the sample to the BMU's weight vector is the quantization error.

Definition 3. *Using the notation from definition 1, the quantization error* **qe** *of an input sample $\mathbf{o}_i \in \mathbb{R}^m, i \in \mathbb{N}$ is given by the distance of the input sample to its BMU of the SOM: $\mathbf{qe} = d(\mathbf{o}_i, bmu(\mathbf{o}_i))$.*

The quantization errors for data that is not anomalous are usually greater than 0 due to the discretization of the SOM. A threshold for the quantization error above which an input sample is classified as anomalous is required. Manual selection of the threshold works but is usually unfeasible for practical applications. It is far more convenient to estimate the threshold from data: the quantization errors of the training data can be seen as a probability distribution and quantiles can be used to retrieve the threshold for the anomaly detection. The quantile can be adjusted and we will use the parameter τ with $\tau \in \mathbb{R}$ and $0.0 \leq \tau \leq 1.0$ within this paper. This can be adjusted to optimize the outcome of the anomaly detection: when labels are present τ can be used to fine tune the anomaly detection score. When the training data is perfect, meaning it contains only normal behavior, no sampling errors, no glitches in the sensors and no noise, then a τ of 1.0 is fine as this results in the maximum error for the threshold. However, training data is never perfect when working with real systems and data might contain a small portion of samples affected by noise and/or other effects. The maximum error might be too large to effectively find anomalies. Setting τ to a value slightly smaller than 1.0 can increase the true positive rate of the anomaly detection at the cost of some false positives, depending on the use case and desired outcome.

2.2 Localization of anomalies

An additional step after the anomaly detection is to calculate the signal or sensor which is most likely to cause the anomaly. Anomaly localization is performed after an anomaly is detected: the observation found to be anomalous is fed through a reverse model to obtain the expected values for the signals. The deviations from the expected values can then be used to identify the signals related to the anomaly. The weight vector of each neuron of the SOM has the same dimensions as the input data and each element of the weight vector contains the value of its corresponding signal. Once an anomaly is detected, the input sample is again mapped to the SOM to retrieve the BMU. Now, the distance of each signal to the weight vector is calculated and the resulting signals and their distances are sorted in a descending order according to their distance. Since real world systems usually provide a large number of signals it is necessary to reduce the number of displayed signals. Therefore only the first n signals are displayed giving plant experts and operators a starting point to locate the anomaly and possible fault in the system and ultimately restore the system's normal working order. For the experiments in this paper we only consider the signal with the largest deviation (n=1).

2.3 SOM trajectory tracking with timed automata

Another way to utilize self-organizing maps for anomaly detection in industrial production processes is to track the trajectory of the working point on top of the SOM. Other works such as [7], [12] and [2] already performed a visual anomaly detection on different processes by tracking the trajectory of the working point on the SOM: the observations are mapped to their corresponding best matching units as soon as they are recorded. Over time, the path or trajectory of the BMU can be observed and deviations from the known path indicate anomalous behavior.

However, these works do not attempt to track the trajectory through the use of a mathematical model and only pursue a visual anomaly detection by plotting the trajectory on top of the SOM's u-matrix. In this section we use discrete timed automata to learn the trajectory during normal production and detect deviations from it afterwards. This provides explicit modeling of time which the SOM is unable to do alone.

Timed automata have proven to be a great tool to learn the normal behavior of a system and detect deviations from it. Discrete events are required to learn an automaton. They often cause mode changes within the system and the timing of these events is an important indicator for the health of the system. Timed automata are used to separate the system's modes and model the transitions and timing between the identified modes.

Discrete events can be directly extracted from changes in the binary control and sensor signals of the system. It is also possible to obtain discrete events through thresholds for real-valued signals such as *temperature* $< 19°C$ [4]. However, setting the thresholds and combinations of conditions for the continuous signals requires expert knowledge which is usually not available for real world automation systems. For unsupervised learning of these automata only binary control signals are used to obtain the discrete events, such as *HeaterOn* = *true*. Algorithms such as the bottom-up timed learning algorithm (BUTLA) [8] work in an unsupervised manner, and do not require additional expert knowledge.

A timed automaton generated by the aforementioned algorithm can be defined as described in Definition 4.

Definition 4. *a timed, probabilistic automaton is a tuple* $A = (S, s_0, \Sigma, T, \delta, P)$, *where*

- S *is a finite set of states where* $s \in S$.
- s_0 *is the initial state which can be given by the systems state at the start of the training.*
- Σ *is the set of discrete events. Events* $a \in \Sigma$ *is linked to the transitions of the automaton.*
- T *is the set of transitions with* $t \in T$ *and* $t = (s, a, s'))$, $s, s' \in S$ *are source and destination state,* $a \in \Sigma$ *is the trigger event of the transition.*
- *The timing constraint* $\delta : T \rightarrow I$ *assigns a time interval to a transition* $t \in T$, *where* I *is a set of time intervals. The time here usually refers to the elapsed time since the last event occurred.*
- P *is a set of probabilities: for each transition* $t \in T$ *probability* $p \in P$ *is calculated.*

The learned automaton can then be used to detect a variety of different classes of anomalies. This can for example be done using the anomaly detection algorithm (ANODA) [8] which can detect the following types of anomalies:

- **Unknown event / Wrong event sequence**: an event occurred which was not observed in the current state.
- **Timing error**: a transition occurred outside of the learned time bounds.
- **State remaining error**: when more time passed than for the latest event and the state is not a final state, then we have a state remaining error.
- **Probability error**: the probabilities of transitions for the new data are calculated and compared to the previously learned probabilities and an error is generated when deviations are too large.

As mentioned before, discrete events are needed to learn an automaton. One way to obtain these from the SOM is to interpret each neuron of the SOM as a binary signal: a neuron is active (or true) when the observation is mapped to it and false otherwise. This mapping can result in a large number of signals, depending on the size of the SOM. This usually leads to a large number of states in the automaton. Also, some neurons might never be activated by the training data leading to an unknown state detection in the automaton. Again, this can lead to a large number of false positives when new data is mapped to neurons which were previously not active but are direct neighbors to neurons previously activated by the training data.

To counteract these effects and ultimately reduce the number of states we group the neurons into a smaller number of clusters. The transitions between the clusters are then learned using a timed automaton.

SOM's tend to keep neurons with similar signal weights closely together, which results in a topographic landscape with valleys, where weights of neighbors are similar, and ridges, where weights of neighbors are not similar. This landscape can be visualized through the aforementioned u-matrix representation. Valleys represent regions where the contained neurons weight vectors are very similar. These valleys are separated by ridges which mark transitions between the different feature spaces. The valleys can be interpreted as stationary process phases while the ridges represent transient process phases [3].

Clustering algorithms from the image processing domain, such as the watershed transformation [9], can be used on the u-matrix representation of a SOM to identify the clusters in a mathematical way. This works analogous to rain falling on top of the u-matrix. The water runs from higher regions to the lower regions and consequently flooding the basins. When the water level gets high enough so two basins merge, a ridge forms which separates them. The watershed transformation dissects the u-matrix into different clusters, separated by the so-called watershed lines. Watershed lines separate the different basins and do not belong to any of the clusters. The implementation used here is the Vincent-Soille watershed algorithm which performs the watershed transformation in a non-recursive manner [15].

Subsequently, the samples of the training data are mapped to the SOM to get the corresponding cluster. The clusters are encoded using a one-hot encoding resulting in a binary vector with one element for each cluster. The value of the active cluster is set to true, while all other values are set to false. The time-stamps of the original samples and the binary vectors are then used to learn an automaton with the aforementioned BUTLA.

3 Experiments

In this section we apply the aforementioned approaches for anomaly detection and anomaly localization to one of our demonstrators. The Genesis demonstrator of the Institute Industrial IT sorts two different materials (conductive and non-conductive) from a magazine into their corresponding target locations (Figure 2). It is portable and uses an air tank to supply all the gripping and storage units. The 4 different modules can switch places and the program for the programmable logic controller (PLC) automatically adjusts for the change in location. A linear drive with a pneumatic gripper transports the materials between the different stations. Five real-valued signals are available from the demonstrator: current, position, speed, acceleration and force. Data samples were taken through an OPC connection with a resolution of 50 milliseconds for a total of 42 production cycles. The first 38 production cycles contain only normal behavior and were used to train the self-organizing map for both experiments shown in this section. Two of the 4 remaining cycles contain anomalous behavior and are used for the anomaly detection.

Fig. 2. The Genesis Demonstrator

3.1 Quantization error anomaly detection and anomaly localization

A self-organizing map is trained over 100 Epochs using a 60x60 square, non toroidal topology on the training data using the Eulidean distance measure. Its unified distance matrix representation can be seen in Figure 3. The training data only contains only normal behavior. The anomaly detection is performed in an unsupervised way.

To estimate the threshold for the anomaly detection, tau was set to 1.0 which gives a value of ~0.274 as threshold for the anomaly detection as shown in Figure 4. The observations of the evaluation data are then mapped to the SOM and the distance is calculated as seen in Figure 5. The observation is marked as an anomaly when the distance is larger than the previously estimated threshold.

The final result of the anomaly detection is shown in Figure 6. Anomalies in this data set are labeled which allows to calculate the quality of the anomaly detection: in this example anomalies were detected with an accuracy of 99.63% and F1 score of 94.34%. Sensitivity was 100% which means all anomalies were identified correctly as true positive. Detailed results are shown in Table 1. Figure 7 shows an excerpt of the anomaly localization of the anomalies detected above. Only the two most likely signals estimated to be the cause of the anomaly are shown. The localization results match the predictions made by the experts on the related signals of the system perfectly. The two anomalous cycles shown in Figure 6 contain the same anomalies: the first part of each anomaly, labeled '1' in the data set, is a jam in the linear drive resulting in a standstill and a higher than usual motor current. The second phase of the anomaly, labeled '2' in the data set, is the linear drive

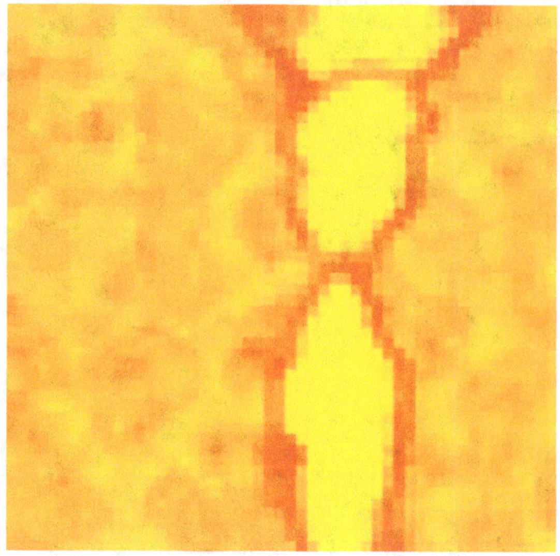

Fig. 3. U-matrix of self-organizing map

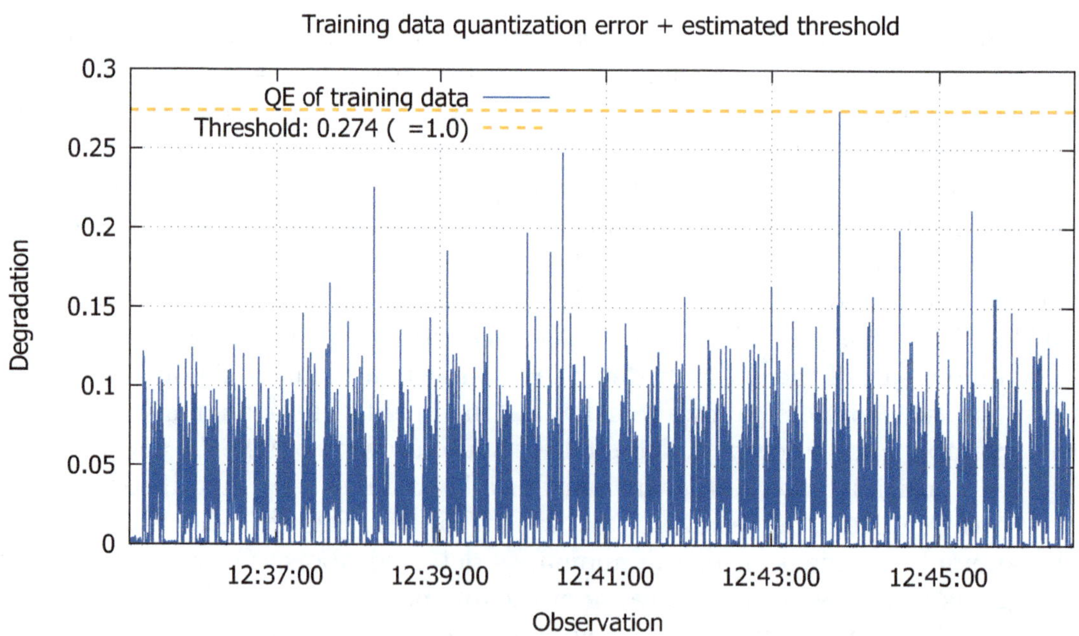

Fig. 4. Quantization error on training data and estimated threshold

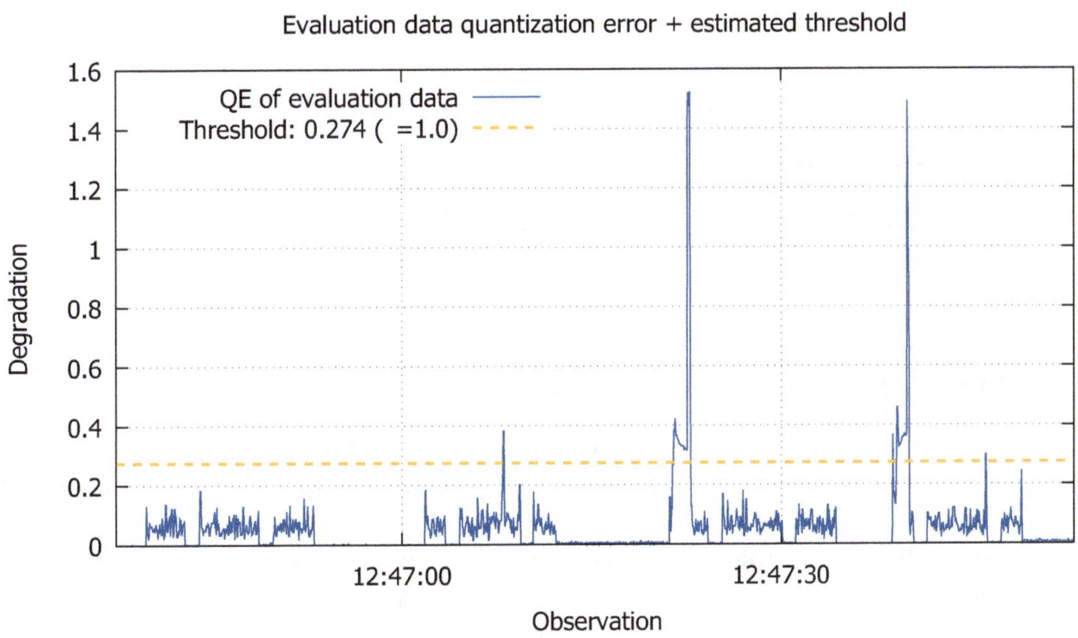

Fig. 5. Quantization error on evaluation data and estimated threshold

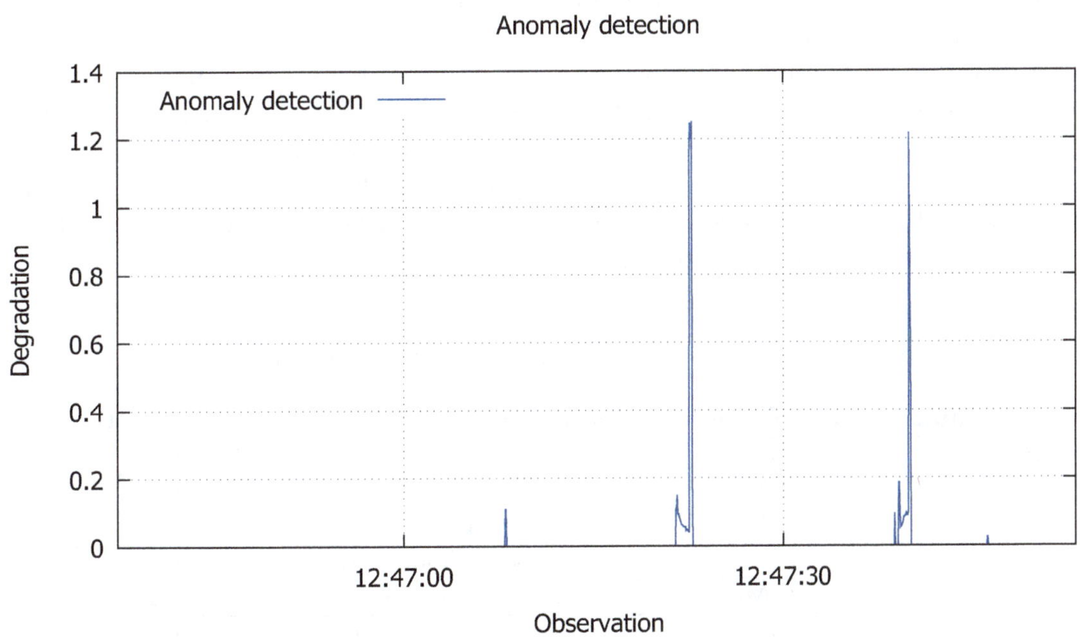

Fig. 6. Final result of the anomaly detection

overcoming the jam, trying to correct its lag error. This results in a much higher than usual speed of the linear drive which is also correctly identified by the anomaly localization.

Table 1. Anomaly detection score

TP	TN	FP	FN	Accuracy	Balanced Accuracy	F1
50	1565	6	0	99.63%	99.81%	94.34%

3.2 Trajectory tracking with automata

Again, a SOM is trained on 38 production cycles containing only normal behavior. The SOM has a size of 60x60 neurons. The resulting u-matrix shown in Figure 8 is then dissected into 6 clusters by the watershed transformation in Figure 9.

Subsequently, the samples of the training data are mapped to the SOM to get the corresponding cluster. The automaton which represents the trajectory across the different clusters during normal operation of the system is learned and shown in Figure 10. The one-hot encoding is easy to read the cluster transitions from the automatons transitions: $C0 = 1; C5 = 0; (5.25 - 10.36)(7.11s)$ describes a transition from cluster 5 to cluster 0 with a timing of 5.25-10.36 seconds after entering the state. The mean time for the transition was 7.11 seconds.

Other encodings than the one-hot can be used so less binary signals are needed for the amount of clusters to describe but they might be harder to read and follow when looking at the automaton.

An example mapping for a single production cycle from the evaluation data is shown in Table 2. Not all states from the state machine can be found in the SOM, as the SOM uses only real-valued signals. Binary signals, such as the storage ejecting material and the gripper closing are not known. The linear drive which provides the real-valued signals does not move during these operations and therefore, these internal states appear in the same cluster of the SOM.

Table 2. Mapping of states to internal state machine and clusters

Automaton state	State machine	State machine description	Cluster
S4	SM0	Idle (standstill)	C0
S5	SM4	Move to storage	C4
S6	SM4	Move to storage (stopping)	C0
S7	SM5	Close gripper (standstill)	C3
S8	SM6	Move to sensor	C2
S9	SM7	Move to target container	C1

```
[...]
20.04.16 12:47:21.438:
MotorData.ActSpeed (0.27)
MotorData.ActCurrent (0.18)
[...]
20.04.16 12:47:22.095:
MotorData.ActCurrent (0.22)
MotorData.ActSpeed (0.21)
[...]
20.04.16 12:47:22.562:
MotorData.ActSpeed (1.50)
MotorData.ActCurrent (0.17)
[...]
20.04.16 12:47:22.703:
MotorData.ActSpeed (1.50)
MotorData.IsForce (0.17)
[...]
[...]
20.04.16 12:47:38.828:
MotorData.IsForce (0.33)
MotorData.ActCurrent (0.14)
[...]
20.04.16 12:47:39.479:
MotorData.ActCurrent (0.21)
MotorData.ActSpeed (0.21)
[...]
20.04.16 12:47:39.998:
MotorData.ActSpeed (1.49)
MotorData.ActCurrent (0.08)
[...]
20.04.16 12:47:40.096:
MotorData.ActSpeed (0.78)
MotorData.IsAcceleration (0.45)
[...]
```

Fig. 7. Excerpt of the anomaly localization algorithm

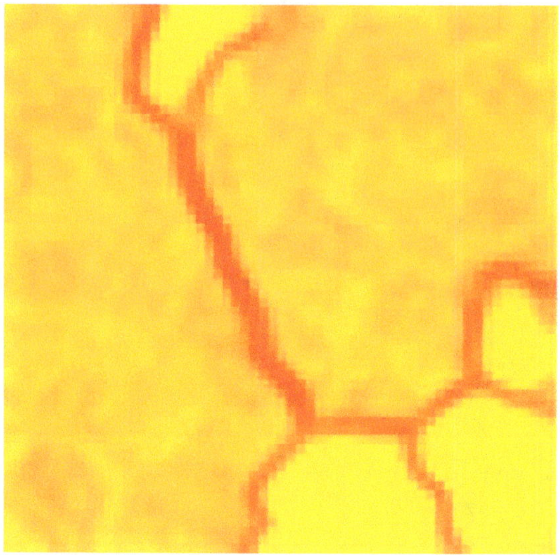

Fig. 8. U-matrix of the SOM

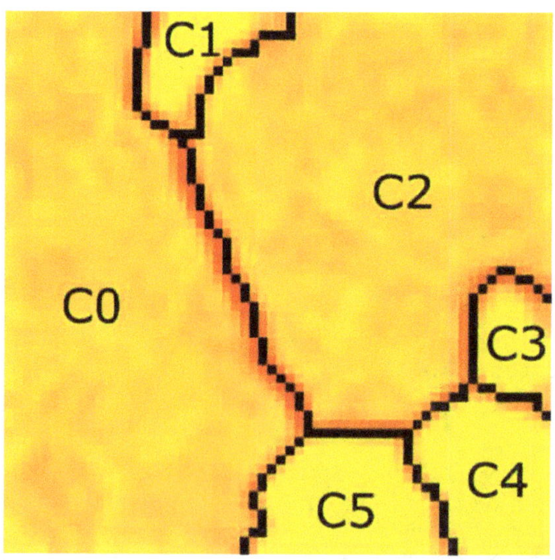

Fig. 9. Clustering after applying watershed transformation

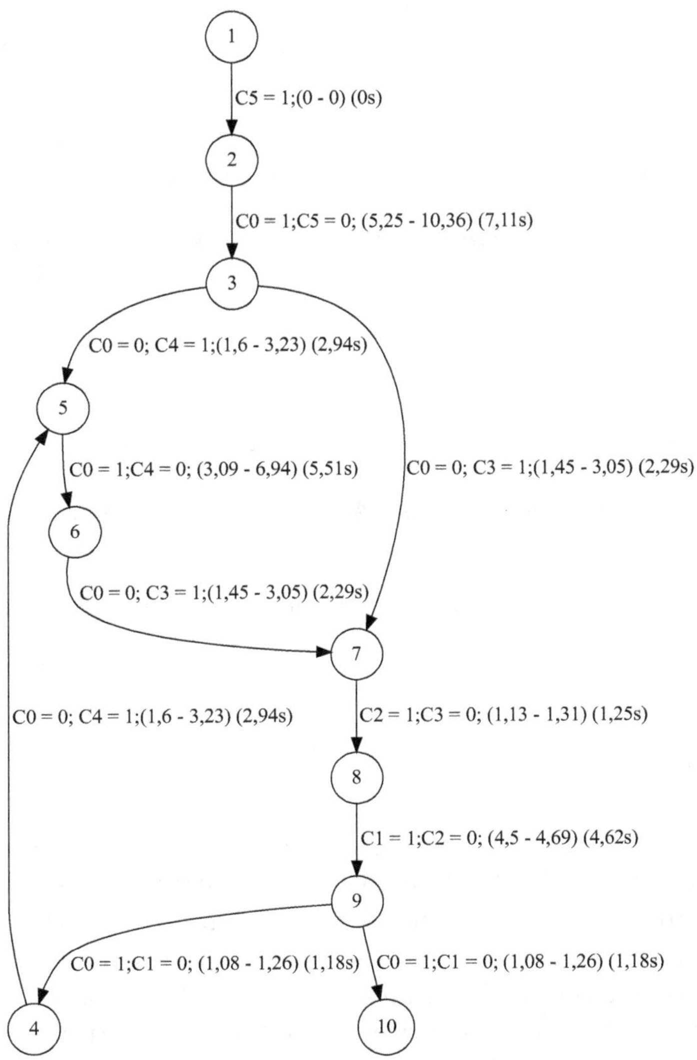

Fig. 10. Automaton tracking the transitions between clusters

Figure 11 shows examples from the output of the ANODA: first, on observation 482 a state remaining error occurs. During the production cycle it took more time to move the gripper to the storage position. This triggers a timing error when the transition finally occurs and normal operation continues. Second, at the end of the data set it was detected that the demonstrator remains in its idle state longer than in the training cycles. Both of these errors are not detectable using the quantization error method presented in the previous section because the SOM itself does not model time in an explicit manner. The automaton adds explicit modeling of time by modeling and tracking the trajectory of the SOM's working point.

```
0481|S5[Normal;]
0482|S5[StateRemainingError;Time is 6.98s but max time in state 5 is 6.94s.]
[...]
0519|S5[StateRemainingError;Time is 8.72s but max time in state 5 is 6.94s.]
0520|S5->S6[TimingError;[8.76s;3.094s to 6.94s with mean 5.50s        ]]
0521|S6[Normal;]
[...]
[...]
01600|S4[Normal;]
01601|S4[StateRemainingError;Time is 3.23s but max time in state 4 is 3.22s.]
[...]
01620|S4[StateRemainingError;Time is 4.12s but max time in state 4 is 3.22s.]
```

Fig. 11. Anomaly detection with the automaton and ANODA

4 Conclusion

This paper presented approaches to data-driven anomaly detection and localization in Cyber-Physical Production Systems. Data provided by the system is used train a self-organizing map to represent the systems normal behavior.

The first option shown in this paper uses the quantization error to detect anomalies in the systems signal domain. Manual adjustment of a threshold above which an anomaly detected is not easy so we estimate the threshold from the data. When an anomaly is found, the SOM is used as a reverse model to compute the differences between expected and actual value of each signal. The signals are sorted from largest to smallest deviation and the first n signals are provided as a starting point for experts to restore the system to a normal working order.

This anomaly detection and localization can be applied to a wide variety of systems and produces good results across the board. However, time is not modeled and deviations in the timing can not be found.

The second option shown in this paper adds modeling of time: discrete timed automata are used to learn the trajectory of the SOM's working point. The automaton keeps track of the timing between the different process phases. This approach detects anomalies in the systems timing and observed sequence, both of which can not be detected by the SOM alone.

The second approach can be extended to use hybrid timed automata instead of discrete timed automata: a separate model is learned for each state of the automaton to model the different stationary process phases and detect anomalies within them [16]. Yet another approach could replace the discrete timed automata with variable order Markov models. The automaton only uses its current state and long term deviations from the trajectory might not be detectable, especially when the automaton contains cycles. In general, higher order Markov models also use a number of previous states to predict the following state and might be able to better deal with cycles and deviations which happen over a long period of time.

Acknowledgments. This project has received funding from the European Unions Horizon 2020 research and innovation programme under grant agreement No. 678867.

References

1. Factories of the Future: MultiAnnual Roadmap for the contractual PPP under HORIZON 2020. European Union, Luxembourg (2013)
2. Alhoniemi, E., Hollmn, J., Simula, O., Vesanto, J.: Process monitoring and modeling using the self-organizing map. Integrated Computer Aided Engineering 6, 3–14 (1999), http://citeseerx.ist.psu.edu/viewdoc/summary?doi=10.1.1.33.573
3. Frey, C.: Monitoring of complex industrial processes based on self-organizing maps and watershed transformations. In: Industrial Technology (ICIT), 2012 IEEE International Conference on (Mar 2012)
4. Henzinger, T.A.: The theory of hybrid automata. In: Proceedings 11th Annual IEEE Symposium on Logic in Computer Science. pp. 278–292 (Jul 1996)
5. Kohonen, T.: The self-organizing map. In: Proceedings of the IEEE. vol. 78 (Sep 1990)
6. Kumpulainen, P., Hätönen, K.: Local anomaly detection for mobile network monitoring. Inf. Sci. 178(20), 3840–3859 (2008)
7. Liukkonen, M., Hiltunen, Y., Laakso, I.: Advanced monitoring and diagnosis of industrial processes. In: 2013 8th EUROSIM Congress on Modelling and Simulation. pp. 112–117 (Sept 2013)
8. Maier, A.: Identification of timed behavior models for diagnosis in production systems. Ph.D. thesis, Paderborn, Univ. (2015)
9. Meyer, F.: Topographic distance and watershed lines. Signal Processing 38(1), 113–125 (jul 1994), http://dx.doi.org/10.1016/0165-1684(94)90060-4
10. Niggemann, O., Lohweg, V.: On the diagnosis of cyber-physical production systems: State-of-the-art and research agenda. In: Twenty-Ninth AAAI Conference on Artificial Intelligence (2015)
11. OPAK-Consortium: Offene engineering-plattform fr autonome, mechatronische automatisierungskomponenten in funktionsorientierter architektur (2017), http://www.opak-projekt.de/
12. Simula, O., Kangas, J.: Process monitoring and visualisation using self-organizing maps (1995)

13. Tian, J., Azarian, M.H., Pecht, M.: Anomaly detection using self-organizing maps-based k-nearest neighbor algorithm. Second European Conference of the Prognostics and Health Management Society 2014 (2014)
14. Ultsch, A., Siemon, H.P.: Kohonen's self-organizing feature maps for exploratory data analysis. In: Proceedings of the International Neural Network Conference (INNC'90 (1990), `http://www.uni-marburg.de/fb12/datenbionik/pdf/pubs/1990/UltschSiemon90`
15. Vincent, L., Soille, P.: Watersheds in digital spaces: an efficient algorithm based on immersion simulations. IEEE Transactions on Pattern Analysis and Machine Intelligence 13(6), 583–598 (Jun 1991)
16. von Birgelen, A., Niggemann, O.: Using self-organizing maps to learn hybrid timed automata in absence of discrete events. In: 22nd IEEE International Conference on Emerging Technologies and Factory Automation (ETFA 2017) (Sep 2017)
17. Yin, H.: The Self-Organizing Maps: Background, Theories, Extensions and Applications, pp. 715–762. Springer Berlin Heidelberg, Berlin, Heidelberg (2008)

A Sampling-Based Method for Robust and Efficient Fault Detection in Industrial Automation Processes

Stefan Windmann and Oliver Niggemann

Fraunhofer IOSB-INA, Lemgo 32657, Germany,
stefan.windmann@iosb-ina.fraunhofer.de

Abstract. In the present work, fault detection in industrial automation processes is investigated. A fault detection method for observable process variables is extended for application cases, where the observations of process variables are noisy. The principle of this method consists in building a probability distribution model and evaluating the likelihood of observations under that model. The probability distribution model is based on a hybrid automaton which takes into account several system modes, i.e. phases with continuous system behaviour. Transitions between the modes are attributed to discrete control events such as on/off signals. The discrete event system composed of system modes and transitions is modeled as a finite state machine. Continuous process behaviour in the particular system modes is modeled with stochastic state space models, which incorporate neural networks. Fault detection is accomplished by evaluation of the underlying probability distribution model with a particle filter. In doing so both the hybrid system model and a linear observation model for noisy observations are taken into account. Experimental results show superior fault detection performance compared to the baseline method for observable process variables. The runtime of the proposed fault detection method has been significantly reduced by parallel implementation on a GPU.

Keywords: fault detection, hybrid systems, filtering

1 Introduction

Reliable fault detection in industrial automation processes allows costs and risks to be reduced by the early detection of faults and problems in the process and by preventing component failures and, in extreme cases, a production stop of the entire plant. Typical goals of fault detection methods for complex and distributed automation systems are the detection of faults, suboptimal energy consumption, or wear (see e.g. [1], [12]). The purpose is the detection of deviations in system behavior from the normal state (for example too high or too low energy consumption of a conveying system). Generally speaking, fault detection methods for industrial automation processes can be divided in model-based methods, signal-based methods and knowledge-based methods (see [4] for an overview). This paper is focused on model-based fault detection. In model-based approaches, the consistency between the measured output of the system and the model-predicted output is checked. In

O. Niggemann and P. Schüller (Eds.), *IMPROVE – Innovative Modelling Approaches for Production Systems to Raise Validatable Efficiency*, Technologien für die intelligente Automation 8, https://doi.org/10.1007/978-3-662-57805-6_5

doing so, significant outliers are detected and can be displayed in a human machine interface. Model-based fault detection methods can be divided in four categories [4]: Deterministic methods, stochastic methods, methods for discrete-events and hybrid systems, and methods for networked and distributed systems.

Complex industrial systems such as complex mechatronic systems, embedded control systems and manufacturing systems are hybrid systems, which are both driven by time-based continuous dynamics and event-driven discrete dynamics. Faults in the underlying discrete event system are straightforwardly detected by comparison of the observed discrete events with events that are predicted with a finite state machine (see e.g. [15]). A lot of research has been conducted with respect to fault detection in continuous system parts. Clustering-based fault detection methods create groups of strongly related objects and objects which do not strongly belong to any cluster [3]. Neural networks and regression-based methods have been used to approximate the functional dependency between continuous process variables and time [15]. Sensor signals are predicted according to this functional dependency and significant deviations of predicted signal values from the observations are reported as faults. Stochastic approaches to fault detection are predominantly based on building a probability distribution model and considering how likely objects are under that model [14]. Statistical tests are used to assess the likelihood of faults (see e.g. [17], [18]). In most of the stochastic approaches, state variables are employed for modeling the temporal transitions of hidden process variables which are related to the observations with an observation model. Common approaches are Kalman filters (e.g. [23], [11]) and particle filters [16]. Hybrid systems require appropriate models, which consider mutual dependencies and interactions between continuous dynamics and discrete-event changes. Hybrid automata are the most common models to represent hybrid systems, which can be utilized to detect and isolate faults (see e.g. [5], [7], [22]). An efficient fault detection method for hybrid industrial automation processes with observable process variables has been proposed in [18].

This paper is an extended version of [21]. The main contribution of the present work is the enhancement of the method proposed in [18] for the case of noisy observations by integration of a particle filter. To increase the efficiency of the particle filter, a parallel implementation has been realized on a graphics processing unit (GPU).

The remaining part of the paper is structured as follows: Section 2 outlines the probabilistic framework for fault detection in hybrid industrial processes. The GPU-based fault detection method is introduced in Section 3. The proposed method is evaluated in a given application scenario (see Section 4). Section 5 gives a conclusion and an outlook to future work.

2 Fault detection with stochastic process models

In this section, the prior work on fault detection with stochastic process models is described. Particularly, the probabilistic framework according to [17] and the underlying hybrid process model adopted from [21] are outlined. The principle of fault detection with stochastic process models is depicted in Fig. 1.

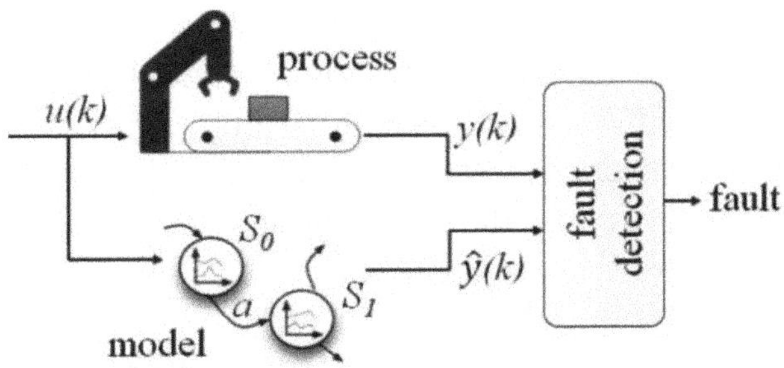

Fig. 1. Probabilistic framework for model-based fault detection.

Measurements $y(k)$ at discrete time k are predicted according to a stochastic process model with input $u(k)$. A common approach to fault detection is the evaluation of the residuals $r(k) = y(k) - \hat{y}(k)$ between measurements $y(k)$ and predictions $\hat{y}(k)$. Methods such as log-likelihood ratios, generalized log-likelihood ratios, sequential probability ratio tests, or cumulative sum test statistics are used in literature to detect changes in the residuals (see [8] for an overview). A related approach is the evaluation of $y(k)$ with respect to the conditional probability density $p(y(k)|y(0)\dots y(k-1))$ (see e.g. [17], [18]). Observations are assumed to be improbable once measurement $y(k)$ at discrete time k drops out of a given confidence interval of the probability density function $p(y(k)|y(0)\dots y(k-1))$. It is shown in [17] that this condition is met once the distribution function $F(y(k))$ of $p(y(k)|y(0)\dots y(k-1))$, i.e. the function

$$F(y(k)) = \int_{\tilde{y}=-\infty}^{y(k)} p(\tilde{y}|y(0)\dots y(k-1))d\tilde{y}$$

with integration variable $\tilde{y}$, is beyond given thresholds, which depend on the width C of the confidence interval ($C \in [0, 1]$):

$$F(y(k)) < \frac{1}{2} - \frac{C}{2} \text{ or } F(y(k)) > \frac{1}{2} + \frac{C}{2}. \tag{1}$$

Measurements beyond a confidence interval of width C are detected as potential faults. In this case, anomaly $f_1(k)$ is reported for time instance k. Furthermore, anomalies with a duration of more than one time instance have to be taken into account as suggested in [18]. Measurements beyond a smaller confidence interval of width C_2 are considered as candidates for such anomalies. If at least a given portion p_d of the last n_d measurements is beyond the confidence interval of C_2, anomaly $f_2(k)$ is reported to indicate small anomalies, which are persistent for more than one time instance. For Gaussian distributions $p(y(k))$, the following relation between the error function $erf(y(k))$ and the distribution function $F(y(k))$ is exploited to determine whether an observation $y(k)$ is located in the allowed confidence interval [17]:

$$F(y(k)) = \frac{1}{2}\left(1 + erf\left(\frac{y(k) - \mu}{\sqrt{2}\sigma}\right)\right). \quad (2)$$

Evaluation of (1) and (2) requires an adequate model of the process behavior, which determines the probability density $p(y(k)|y(0)\ldots y(k-1))$. A large part of stochastic fault detection methods (see e.g. [17], [23], [11], [16]) is based on the following discrete time nonlinear state space model:

$$y(k) = h(\mathbf{x}(k), u(k)) + v(k) \quad (3)$$

$$\mathbf{x}(k) = \mathbf{f}_\Theta(\mathbf{x}(k-1), u(k)) + \mathbf{w}(k) \quad (4)$$

where $\mathbf{f}_\Theta$ and h are the state and measurement dynamic functions, respectively. The two equations consider uncertainties in the form of additive terms for measurement noise $v(k)$ and state transition noise $\mathbf{w}(k)$.

The continuous system behavior is modeled with a time-variant state vector $\mathbf{x}(k)$. Estimation of $p(y(k)|y(0)\ldots y(k-1))$ with the underlying state-space model (3) and (4) requires a recursion with the following steps [13]:

1. $p(\mathbf{x}(k)|y(0)\ldots y(k-1)) = \int \{p_\Theta(\mathbf{x}(k)|\mathbf{x}(k-1)) \cdot p(\mathbf{x}(k-1)|y(0)\ldots y(k-1))\}\, d\mathbf{x}(k-1)$
2. $p(y(k)|y(0)\ldots y(k-1)) = \int \{p(y(k)|\mathbf{x}(k))p(\mathbf{x}(k)|y(0)\ldots y(k-1))\}\, d\mathbf{x}(k)$
3. $p(\mathbf{x}(k)|y(0)\ldots y(k)) \propto p(y(k)|\mathbf{x}(k))p(\mathbf{x}(k)|(y(0)\ldots y(k-1))$ (5)

where $p(\mathbf{x}(k)|y(0)\ldots y(\tilde{k}))$ denotes the conditional probability density of state vector $\mathbf{x}(k)$ given the measurements $y(k)$ for time instances $k = 0\ldots\tilde{k}$. Note, that the probability densities

$$p(y(k)|\mathbf{x}(k)) \quad \text{and} \quad p_\Theta(\mathbf{x}(k)|\mathbf{x}(k-1)) \quad (6)$$

are given by (3) and (4). Hence, measurement model (3) and state model (4) form the basis of the proposed fault detection method.

An appropriate measurement model for application cases with noisy observations is introduced in Section 3, while the state model for hybrid processes has been adopted from [21]. The underlying state model is based on a timed automaton with discrete events a assigned to the transitions between system modes $S_0, S_1\ldots$ (see Fig. 1). The time-dependent continuous process behavior in each mode is incorporated by models $\Theta_{S_0}, \Theta_{S_1}, \ldots$. In this work, the following formalism of timed hybrid automata is adopted from [9]:

Definition 1. *A* ***Timed Hybrid Automaton*** *is a 5-tuple* $A = (S, \Sigma, T, \delta, \mathbf{\Theta})$, *where*

- S *is a finite set of states. Each state* $s \in S$ *is a bijective function* $s = f_c(\mathbf{d})$ *of vector* $\mathbf{d} = (d_1, d_2, ..., d_{|\Sigma|})^T$, *which indicates for each IO value, whether it is active* ($d_i = 1$) *or inactive* ($d_i = 0$).
- Σ *is the alphabet, the set of events* a.
- $T \subseteq S \times \Sigma \times S$ *gives the set of transitions. E.g. for a transition* $\langle s, a, s' \rangle$, $s, s' \in S$ *are the source and destination state and,* $a \in \Sigma$ *is the trigger event.*
- *A transition timing constraint* $\delta : T \rightarrow I$, *where* I *is a set of intervals.* δ *refers to the time spent since the last event occurred and is expressed as a time range.*
- $\mathbf{\Theta} = \{\Theta_s\}$ *denotes continuous process behavior, which is assigned to the particular states* $s \in S$ *of the timed automaton.*

Discrete state s, i.e. the system mode, is defined over the active/inactive IO values. The starting point is the system mode corresponding to the actual system's configuration. Model learning for the discrete system part, i.e. the underlying automaton, is accomplished using the OTALA algorithm [9].

A Markov model of order l is assumed for the continuous process behavior Θ_s in the particular system states s. This means, process variables $x(k)$ are assumed to depend on the previous process variables $x(k-l) \ldots x(k-1)$. This is taken into account by extending the state vector $\mathbf{x}(k)$ with the respective process variables:

$$\mathbf{x}(k) = [x(k-l) \ldots x(k)]^T. \tag{7}$$

The initial probability density

$$p_0(\mathbf{x}(0)) = \mathcal{N}(\mu_0, \sigma_0)^l \tag{8}$$

of state vector $\mathbf{x}(0)$ at time instance $k = 0$ is obtained from n samples of noise-free training data $x(k)$, which are acquired in the initial system state s_0:

$$\mu_0 = \frac{1}{n} \sum_{k=0}^{n-1} x(k), \tag{9}$$

$$\sigma_0 = \sqrt{\frac{1}{n-1} \sum_{k=0}^{n-1} (x(k) - \mu_0)^2}. \tag{10}$$

State equation (4) is assumed to depend on system state s:

$$\mathbf{x}(k) = \mathbf{f}_{\Theta_s}(\mathbf{x}(k-1), u(k)) + \mathbf{w}(k). \tag{11}$$

It is worth noting that the dependency on s has no impact on recursion (5) unless Θ_s has to be updated before each recursion step with respect to the current system mode s. The underlying Markov model of order l is taken into account by the state space model

$$\mathbf{f}_{\Theta_s}(\mathbf{x}(k-1), u(k)) = \begin{bmatrix} x(k-l+1) \\ \dots \\ x(k-1) \\ f_{\Theta_s}(\mathbf{x}(k-1), u(k)) \end{bmatrix} \tag{12}$$

$$\mathbf{w}(k) = [0 \dots 0, w(k)]^T. \tag{13}$$

The process noise sequence $w(k), k = 0 \dots n-1$, is assumed to be independent and identically distributed (i.i.d) with standard deviation σ_s. Model parameters Θ_s and standard deviation

$$\sigma_s = \frac{1}{n-1}\sqrt{\sum_{k=0}^{n-1}(x(k) - f_{\Theta_s}(\mathbf{x}(k-1), u(k)))^2} \tag{14}$$

are estimated from n samples of noise-free training data with observable state variables $\mathbf{x}(k)$.

In this work, dependency $f_{\Theta_s}(\mathbf{x}(k-1), u(k))$ is modeled as a feed-forward neural network (multi-layer perceptron). For the evaluated application case, this approach has shown to be superior compared to linear regression models with respect to modeling accuracy (see [20]). A back propagation algorithm is employed to train the neural network [2], which is shown in Fig. 2. The neural network consists of three layers.

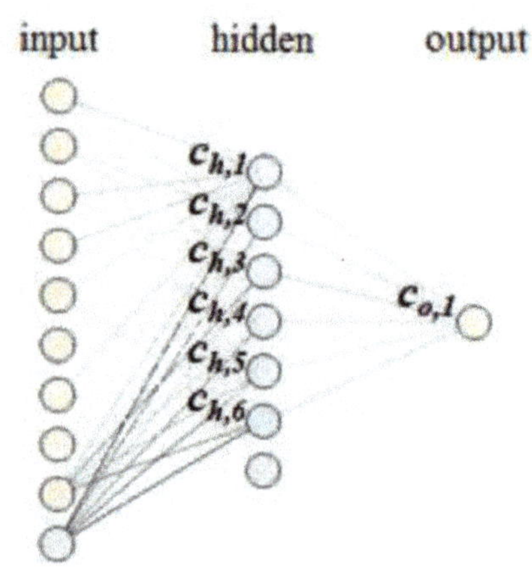

Fig. 2. Neural network with one hidden layer.

The input layer passes the input vector

$$\mathbf{u}(k) = [x(k-l) \dots x(k-1), u(k), 1] \tag{15}$$

to the particular nodes of the hidden layer. The output $x_{h,i}$ of node i of the hidden layer is given by the following equation [2]:

$$x_{h,i} = sig(\mathbf{c_{h,i}}\mathbf{u}(k)) \tag{16}$$

where $sig(x)$ denotes the sigmoid activation function

$$sig(x) = \frac{1}{1 + exp(-x)} \tag{17}$$

for the hidden nodes.

The overall output of the neural network is calculated as linear combination of the components of $\mathbf{x_h} = [x_{h,i}]$:

$$f_\Theta(x(k-l)\ldots x(k-1), u(k)) = \mathbf{c_{o,1}}\mathbf{x_h}. \tag{18}$$

In the hidden layer, a neuron with no input is used to take account for constant offsets. Model parameters $\Theta = (\mathbf{c_{h,i}}, \mathbf{c_{o,1}})$, i.e. the coefficients of hidden layer and output layer, are obtained by back-propagation.

3 Fault detection for application cases with noisy measurements

The proposed fault detection method for application cases with noisy measurements is introduced in this section.

Fault detection with stochastic process models is based on the probability density $p(y(k)|y(0)\ldots y(k-1))$ as described in the previous section. The evaluation of $p(y(k)|y(0)\ldots y(k-1))$ is straightforward for observable process variables [21]. However, in the application cases of this paper, state variables $x(k)$ are not directly observable due to additive noise. This is taken into account by the following measurement model, which is a special case of (3):

$$y(k) = x(k) + v(k) \quad \text{with} \quad v(k) \propto \mathcal{N}(0, \sigma_y). \tag{19}$$

Noise level σ_y is obtained at the beginning of the observation phase from observations $y(k)$ in an initial time slot $k = 0\ldots k_{init}$, where $x(k) = 0$ is assumed.

Fault detection for application cases with noisy measurements involves the evaluation of (5), which in turn requires approximations of the probability densities

$$p(\mathbf{x}(k)|y(0)\ldots y(k-1)) \tag{20}$$

and

$$p(\mathbf{x}(k)|y(0)\ldots y(k)) \tag{21}$$

as stated above. In the present work, particles are employed for this purpose. This approximation is known to be more precise but computationally more intensive compared to the use of the second order statistics that are applied in Kalman Filters. Approximations of the respective probability densities by means of particles are outlined in Section 3.1. The particle filter based fault detection method is introduced in Section 3.2. Computational efficiency is increased by a parallel implementation on a graphics processing unit, which is detailed in Section 3.3.

3.1 Probability density models

The probability densities $p(\mathbf{x}(k)|y(0)\dots y(\kappa)), \kappa \in \{k-1, k\}$, are approximated by the weighted sums [13]

$$p(\mathbf{x}(k)|y(0)\dots y(\kappa)) \approx \sum_{i=0}^{n_p-1} w_i(k)\delta(\mathbf{x}(k) - \mathbf{x}_i(k)) \tag{22}$$

where the n_p particles $\mathbf{x}_i(k)$, $i = 0 \dots n_p - 1$, are sampled from an approximation $q(\mathbf{x}(k)|\mathbf{x}_i(k-1), y(k))$ of the optimal importance density [13]

$$q_{opt}(\mathbf{x}(k)|\mathbf{x}_i(k-1), y(k)) = p(\mathbf{x}(k)|\mathbf{x}_i(k-1), y(k)). \tag{23}$$

In (23), $\mathbf{x}_i(k-1)$ denotes the *ith* particle at time instance $k-1$. Weights $w_i(k)$ are obtained by the following equation [13]:

$$w_i(k) \propto w_i(k-1)\frac{p(y(k)|\mathbf{x}_i(k))p(\mathbf{x}_i(k)|\mathbf{x}_i(k-1))}{q_{opt}(\mathbf{x}_i(k)|\mathbf{x}_i(k-1), y(k))} \tag{24}$$

$$\text{with} \quad \sum_{i=0}^{n_p-1} w_i(k)) = 1. \tag{25}$$

Sampling from the optimal importance density is not straightforward because $q_{opt}(\mathbf{x}(k)|\mathbf{x}_i(k-1), y(k))$ is in the general case not a Gaussian distribution. The most popular suboptimal choice is the transitional prior

$$q(\mathbf{x}(k)|\mathbf{x}_i(k-1), y(k)) = p(\mathbf{x}(k)|\mathbf{x}_i(k-1)). \tag{26}$$

However, in system modes with high standard deviation σ_s of the state transition noise, many particles may be generated that are unlikely with respect to the measurement model. Therefore, the following approximation for the optimal importance density is employed in this work:

$$\begin{aligned} q(\mathbf{x}(k)|\mathbf{x}_i(k-1), y(k)) &= p(\mathbf{x}(k)|\mathbf{x}_i(k-1)), \text{ for } \sigma_s < \sigma_y \\ q(\mathbf{x}(k)|\mathbf{x}_i(k-1), y(k)) &= p(\mathbf{y}(k)|\mathbf{x}_i(k-1)), \text{ else.} \end{aligned} \tag{27}$$

Substitution of (27) in (24) yields the respective equations for weight update:

$$\begin{aligned} w_i(k) &\propto w_i(k-1)p(y(k)|\mathbf{x}_i(k)), \text{ for } \sigma_s < \sigma_y \\ w_i(k) &\propto w_i(k-1)p(\mathbf{x}_i(k)|\mathbf{x}_i(k-1)), \text{ else.} \end{aligned} \tag{28}$$

3.2 Particle filter based fault detection

The proposed fault detection method (Algorithm 1) is shown in Fig. 3. The approach is based on a fault detection method for observable process variables, which is described in Appendix A [19]. A particle filter, which uses the approximations

Given:
(1) Hybrid Automaton $A = (S, \Sigma, T, \delta, \mathbf{\Theta})$
(2) Standard deviation σ_s of the process noise
(3) Standard deviation σ_y of the measurement noise
(4) Number of observations n
(5) Observations $\mathbf{o}(k) = (t(k), \mathbf{d}(k), u(k), y(k)), k = 0 \dots n-1$
(6) Number of particles n_p
(7) Confidence interval C
Results: Detected faults (if there exists one), otherwise "OK"

$s(0) = s_0 = f_c(\mathbf{d}(0))$
for $i = 0 \dots n_p - 1$ **do** $w_i = 1/n_p$, draw $\mathbf{x}_i(0) \propto p_0(\mathbf{x}(0))$
for $k = 1 \dots n-1$ **do**
  $a(k) = generateEvent(\mathbf{d}(k-1), \mathbf{d}(k))$
  $T' = \{e = (s(k-1), a(k), *) \in T\}$
  if $T' = \phi$ **then return** fault: unknown event
  while $isNotEmpty(T')$ **do**
    if $t(k) \in \delta(e = (s(k-1), a(k), s_{new}))$ **then**
      $s(k) := s_{new}$
    else if $t < min(\delta(e))$ **then return** fault: event too early
    else if $t > max(\delta(e))$ **then return** fault: event too late
    for $i = 0$ to $n_p - 1$ **do**
      if $\sigma_s < \sigma_y$
        Draw $\mathbf{x}_i(k) \propto p_{\Theta_{s(k)}}(\mathbf{x}(k)|\mathbf{x}_i(k-1))$
        $w_i(k) = w_i(k-1)p(y_k|\mathbf{x}_i(k))$
      else
        Draw $\mathbf{x}_i(k) \propto p(y(k)|\mathbf{x}(k))$
        $w_i(k) = w_i(k-1)p(\mathbf{x}_i(k)|\mathbf{x}_i(k-1))$
      end if
    end for
    for i $= 0$ to $n_p - 1$ **do**
      $w_i(k) = w_i(k) / \sum_{i=0}^{n_p-1} w_i(k)$ //Normalization
    end for
    Metropolis Resampling (Algorithm 4 in Appendix B)
    $\mu(k) = \sum_{i=0}^{n_p-1} w_i(k-1) f_{\Theta_{s(k)}}(\mathbf{x}_i(k-1), u(k))$
    $\sigma^2(k) = \sigma_y^2 + \sigma_s^2 + \sum_{i=0}^{n_p-1} w_i(k-1)(f_{\Theta_{s(k)}}(\mathbf{x}_i(k-1), u(k)) - \mu(k))^2$
    $F(y(k)) = \frac{1}{2}\left(1 + erf\left(\frac{y(k)-\mu(k)}{\sqrt{2}\sigma(k)}\right)\right)$
    if $F(y(k)) < 1/2 - C/2$ **then return** fault: signal drop
    if $F(y(k)) > 1/2 + C/2$ **then return** fault: signal jump
  end while
end for

Fig. 3. Algorithm 1 Particle filter based fault detection

introduced in the previous Section 3.1, has been incorporated in the underlying fault detection method.

The fault detection method evaluates at each time instance $k = 0 \dots n-1$ the measurement vector

$$\mathbf{o}(k) = (t(k), \mathbf{d}(k), u(k), y(k)), \tag{29}$$

which comprises a time stamp $t(k)$, discrete measurements $\mathbf{d}(k)$, a continuous input value $u(k)$ and a continuous measurement $y(k)$ (input (5) of Algorithm 1). In doing so, the hybrid system model A introduced in Section 2 (input (1)) is used to evaluate the likelihood of possible faults. The standard deviations σ_s and σ_y of process noise and measurement noise are given as input (2) and input (3), respectively. The number n of processed observations is given as input (4). Further parameters of Algorithm 1 are the number n_p of particles used to approximate the respective probability densities (input (6)), and the width of the confidence interval C, which is introduced in (1) (input (7)).

The initial state $s_0 = f_c(\mathbf{d}(0))$ of the discrete event system is given by the active/inactive IO values $\mathbf{d}(0)$ at the beginning of operation. Furthermore, the distribution of the continous state variable $\mathbf{x}(0)$ has to be represented by particles at the beginning of operation. Therefore, particles $\mathbf{x}_i(0)$ are drawn from $p_0(\mathbf{x}(0))$.

At each time instance, observation $\mathbf{o}(k)$ is processed. Fault detection in the discrete event system is conducted according to [15]. Events $a(k)$ are determined from the difference between discrete IO signals $\mathbf{d}(k)$ and $\mathbf{d}(k-1)$. Time $t(k)$ denotes the continuous time stamp of the kth sample, $\mathbf{d}(k)$ the discrete IO values and $y(k)$ a value-continuous measurement at discrete time k. Transitions between system modes $s(k-1)$ and $s(k)$ at time instances $k-1$ and k are triggered by discrete events $a(k)$. Unknown events and timing errors of events are detected in lines (10) and (11) of Algorithm 1, respectively.

Fault detection in the continuous system part is accomplished in lines (12)-(29). The distribution of the state vector is updated in lines (13)-(23) according to the particle filter equations (26)-(28). Condition (25) is asserted by the normalization in line (21)-(23). A well-known challenge of particle filters is the degeneracy phenomenon, i.e. the effect that all but one particle i have negligible weights $w_i(k)$ after some iterations [13]. This problem is solved by an additional resampling step as proposed in [13]. The resampling step in line (24) of the proposed fault detection method is accomplished using Metropolis resampling, which has been suggested by the authors of [10] for parallel resampling (see Algorithm 4 in Appendix B). Fault detection in lines (25)-(29) is conducted according to eqs. (1) and (2) with parameters

$$\begin{aligned}\mu(k) &= E[y(k)|y(0)\dots y(k-1)]\\ &= E[f_{\Theta_{s(k)}}(\mathbf{x}(k-1), u(k)) + w(k) + v(k)|y(0)\dots y(k-1)]\\ &= E[f_{\Theta_{s(k)}}(\mathbf{x}(k-1), u(k))|y(0)\dots y(k-1)]\\ &= \sum_{i=0}^{n_p-1} w_i(k-1) f_{\Theta_{s(k)}}(\mathbf{x}_i(k-1), u(k))\end{aligned} \tag{30}$$

and

$$\sigma^2(k) = Var[y(k)|y(0)\dots y(k-1)]$$
$$= Var[f_{\Theta_{s(k)}}(\mathbf{x}(k-1), u(k))|y(0)\dots y(k-1)] + \sigma_y^2 + \sigma_s^2,$$

with

$$Var[f_{\Theta_{s(k)}}(\mathbf{x}(k-1), u(k))|y(0)\dots y(k-1)]$$
$$=E[(f_{\Theta_{s(k)}}(\mathbf{x}(k-1), u(k)) - \mu(k))^2|y(0)\dots y(k-1)]$$
$$=\sum_{i=0}^{n_p} w_i(k-1)(f_{\Theta_{s(k)}}(\mathbf{x}_i(k-1), u(k)) - \mu(k))^2. \quad (31)$$

3.3 Parallel implementation

The time consuming operations of the proposed fault detection method are lines (14), (15), (17), (18), (22), (24), (25) and (26) in Algorithm 1. Hence, lines (12) - (26) of Algorithm 1 have been parallelized on a GPU using a CUDA C implementation. Parallelization of the respective steps is accomplished as follows:

1. Particles are drawn in parallel from $p_{\Theta_{s(k)}}(\mathbf{x}(k)|\mathbf{x}_i(k-1)$ and $p_{\Theta_{s(k)}}(y(k)|\mathbf{x}(k))$ in lines (14) and (17), respectively.
2. Weights are multiplied in parallel with the likelihoods $p(y(k)|\mathbf{x}_i(k))$ and $p(\mathbf{x}_i(k)|\mathbf{x}_i(k-1))$ in lines (15) and (18), respectively.
3. The sum of weights in line (22) and the sums in lines (25) and (26) are obtained by application of the parallel reduction algorithm [6], respectively. The parallel reduction algorithm is a bottom up approach, which computes the sum of some given values, e.g. of the particle weights, in several iterations. The given values are divided in each iteration into pairs and the sum of each pair is computed in parallel, so that the number of values is approximately halved after each iteration. Hence, the sum of n_p values is computed in asymptotic time $O(log\ n_p)$ using parallel reduction.
4. For resampling in line (24), parallel Metropolis Resampling [10] is used (Algorithm 4 in Appendix B).

The parallel implementation of the proposed fault detection method needs a synchronization point before the application of the parallel reduction algorithms, i.e. after the drawing of particles and the weight updates. After this synchronisation point, the values of all particles have to be updated to compute sums. Further synchronization points are required after each iteration of the parallel reduction algorithms.

4 Evaluation and Discussion

The proposed fault detection method has been evaluated in the research conveying system depicted in Fig. 4. The system comprises five conveyor belts and seven drives, which are used to move objects between four storage positions on two levels (five drives for horizontal movements and two drives for vertical movements). A power meter captures continuously the overall power consumption. Further sensor information is collected in the programmable logic control (PLC) of the storage system.

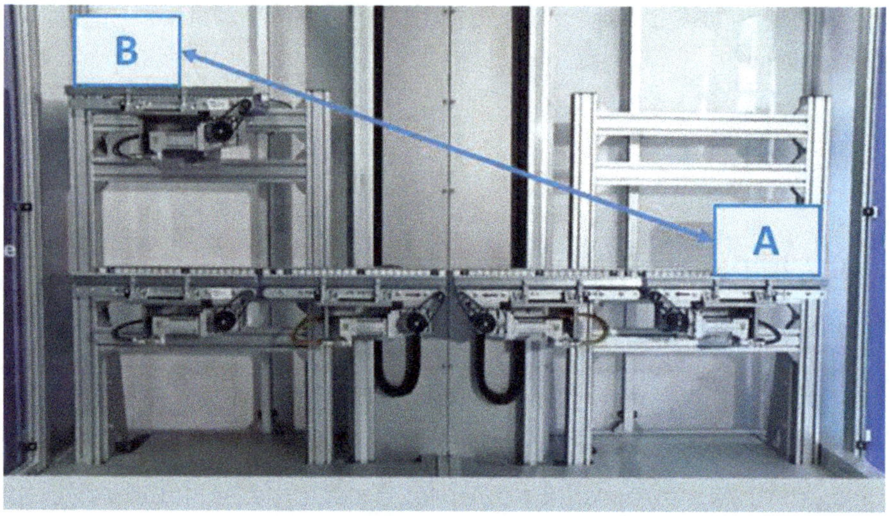

Fig. 4. The research conveying system.

For evaluation, process cycles have been recorded, in which an object has been moved between two positions A and B (see Fig. 4). Fault detetection results for these process cycles are given in Section 4.1. Computational capabilities of the proposed fault detection method have been evaluated in Section 4.2.

4.1 Fault detection results

Fault detection has been accomplished using the particle filter based method introduced in Section 3 (algorithm 1 in Fig. 3) with $n_p = 128$ particles and parameter $n_b = 50$ of the resampling method described in [10]. In the first step, hybrid system models according to Section 2 have been learnt from 40 cycles of training data. Fault detection has been conducted on 10 distinct test cycles with different noise levels. Characteristic faults for increased or decreased motor torques were injected at randomly selected positions into the measurements of the power consumption for the 10 test cyles. The offset of an injected fault (i.e. the injected signal jump or signal drop) was fixed at a value of $\frac{1}{3}\hat{y}$, which depends on the peak power $\hat{y}$ in a particular test cycle. Noise with different noise levels has been artificially added to the test cycles after fault injection.

In order to evaluate the results, ratios of the numbers of correctly and incorrectly predicted samples were computed. Based on the counts of true positive (TP), false positive (FP), false negative (FN) and true negative (TN) samples, the F-measure

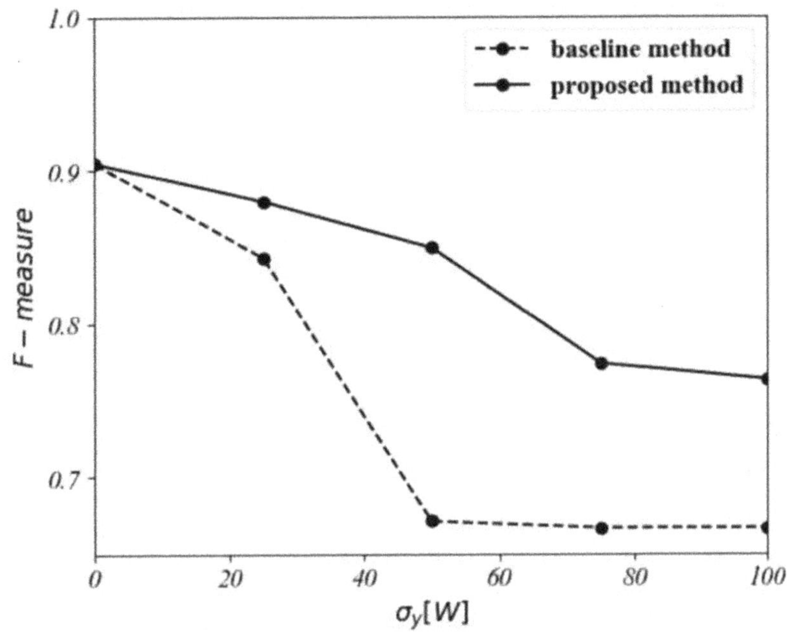

Fig. 5. F-measure for the proposed fault detection method and the baseline method as function of the noise level σ_y.

was calculated as performance metric:

$$\text{F-measure} = 2 \cdot \frac{\text{Sensitivity} \cdot \text{Precision}}{\text{Sensitivity} + \text{Precision}} \tag{32}$$

with

$$\text{Sensitivity} = \frac{\text{TP}}{\text{TP} + \text{FN}} \quad \text{and Precision} = \frac{\text{TP}}{\text{TP} + \text{FP}}.$$

The results for five different noise levels ($\sigma_y = 0, 0.015\hat{y}, \ldots, 0.06\hat{y}$) have been averaged over the 10 test cycles and 25 randomly selected positions for each test cycle. The resulting F-measure for the proposed fault detection method with parameter $C = 0.999$ of the confidence interval is shown in Fig. 5 (solid line). Results of the baseline method [19] (see Fig. 8 in appendix A), which has been developed for automation processes with observable process variables are shown for comparison (dashed line). Both methods are observed to produce the same results for process data without artificial noise. However, the proposed particle filter yields significantly higher F measures for increased noise levels compared to the baseline method. For $n_p = 128$ particles and noise levels between $\sigma_y = 0.015\hat{y}$ and $\sigma_y = 0.06\hat{y}$, an average F-measure of 0.83 is achieved, while the baseline method yields an average F-measure of 0.75. The results confirm the assumption that the influence of additive noise has to be considered in the process model. The proposed fault detection method, which incorporates an appropriate model for additive noise,

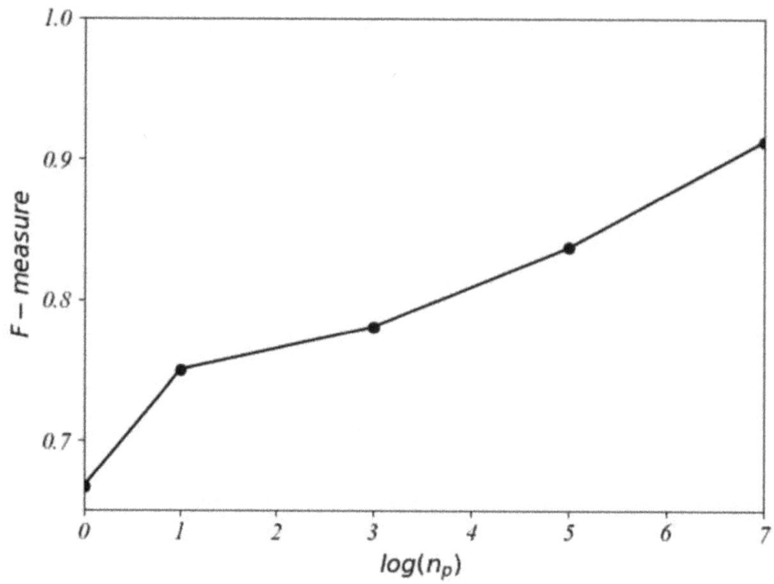

Fig. 6. Impact of the number of particles n_p on the F-measure at noise level $\sigma_y = 100W$.

is shown to be more stable under the condition of increasing noise levels compared to the baseline method.

Fault detection results depend on the number n_p of particles used in the proposed fault detection method. The impact of parameter n_p on the F-measure is shown in Fig. 6 for the first test cycle and noise level $\sigma_y = 100W$. Fig. 6 shows that the F-measure increases with n_p. Particularly, a significant improvement between parameter values $n_p = 1$ ($log_2(n_p) = 0$) and $n_p = 2$ ($log_2(n_p) = 1$) occurs. In informal experiments, the influence of n_p has been observed to be less significant for lower noise levels. However, $n_p = 128$ has turned out to be a good choice to obtain stable results for all investigated noise levels.

4.2 Runtime analysis

The runtime t_{run} of the proposed fault detection method has been evaluated on the 10 test cycles introduced in Section 4.1 with 65000 observations in total. The GPU-based fault detection method has been implemented in CUDA C and has been evaluated on a NVIDIA Quadro K2200 with 640 cores, 4GB GDDR5 memory and a storage band width of 80 GB/s. For comparison, a C implementation of the fault detection method has been evaluated on a single core of an Intel Xeon CPU $E3$-1271 with a clock rate of $3.6GHz$, which employes a RAM with a capacity of 16GB.

Fig. 7 shows the runtime t_{run} of the proposed fault detection method on the GPU (solid line) and on the CPU (dashed line). For more than 16 particles, the GPU-based fault detection method has been observed to outperform the CPU-

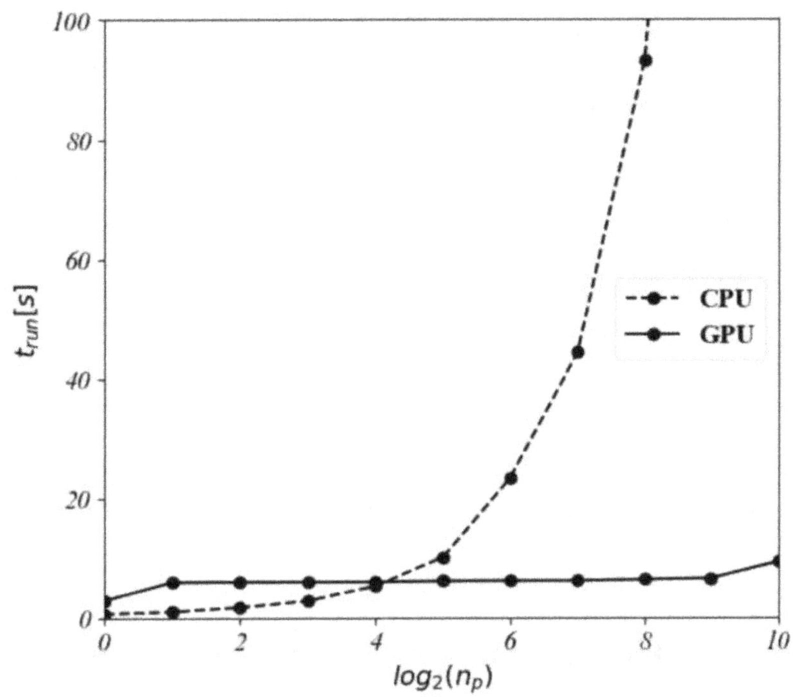

Fig. 7. Runtime t_{run} on GPU and CPU as function of the number of particles n_p.

based implementation. Nevertheless, it should be noted that there is a significant increase in runtime, once the number of particles is changed from $n_p = 512$ ($log_2(n_p) = 9$) to $n_p = 1024$ ($log_2(n_p) = 10$). This is due to the limited number of cores in the GPU, which is not able to process more than 640 particles in parallel. On average, the runtime of the particle filter has been reduced from a value of $t_{run} = 84.19s$ on the CPU to a value of $t_{run} = 6.25s$ on the GPU.

5 Conclusion

In this paper, fault detection for strictly continuous processes has been investigated for applications, in which process variables are not directly observable due to additive noise. The proposed method incorporates a particle filter with switching neural networks in a fault detection method, which has originally been developed for industrial processes with observable process variables. For noise-free test cycles of the considered conveying application, equal results were observed for the proposed fault detection method and the baseline method, which is based on the assumption of noise-free observations. For noise levels between 1.5% and 6% of the energy peak in the respective test cycles, the overall F-measure could be improved from 0.75 to 0.83 by application of the proposed fault detection method compared to the baseline method. The runtime of the proposed fault detection method has been reduced by parallel implementation on a GPU. In doing so, the average runtime of the proposed fault detection method has been decreased for the considered test cycles with 65000 observations from $84.19s$ on a core of an Intel Xeon CPU with $3.6GHz$ to $6.25s$ on a NVIDIA Quadro $K2200$ GPU with 640 cores.

In the present work, application cases with linear measurement models have been investigated. However, extension of the proposed method for application cases with non-linear measurement models is straightforward due to the particle-based approach. Furthermore, the focus has been set on the detection of short-time anomalies, which are typical for blockades or similar error types. An open research question is the application of the proposed fault detection method for processes with gradual degradations by analyzing the probability of the observations for an extended time frame.

Acknowledgments. This project has received funding from the European Union's Horizon 2020 research and innovation programme under grant agreement No. 678867.

Appendix A: Fault detection for observable process variables

In the following section, fault detection for observable continuous process variables is considered. The described approach [20] is used as baseline for the fault detection method that has been proposed in this paper.

Fault detection in the discrete event system is essentially the same for both the proposed fault detection method (lines (1) and (3)-(11) of algorithm 1) and the baseline method (lines (1)-(10) of algorithm 3). In [20], fault detection in the continuous system part (lines (12)-(16) of algorithm 3) is accomplished with the simplified observation model

$$y(k) = x(k), \tag{33}$$

i.e. the assumption of observable continuous state variables $x(k)$. Under this assumption, the stochastic process model (6) is simplified to the single probability density

$$p(y(k)|y(k(0)\dots y(k-1)). \tag{34}$$

Therefore, fault detection is essentially reduced to the evaluation of the distribution function $F(y(k))$ in (1), i.e. the integral of (34), for given measurements $y(0)\dots y(k)$.

Appendix B: Metropolis Resampling

The proposed fault detection method (Algorithm 1) uses a parallel resampling method [10], which is detailed in Algorithm 4.

Given:
(1) Hybrid Automaton $A = (S, \Sigma, T, \delta, \mathbf{\Theta})$
(2) Observations $\mathbf{o}(k) = (t(k), \mathbf{d}(k), u(k), y(k)), k = 0 \ldots n-1$
Results: Detected faults (if there exists one), otherwise "OK"

$s(0) = s_0 = f_c(\mathbf{d}(0))$
for $k = 0 \ldots n-1$ **do**
  $a(k) = generateEvent(\mathbf{d}(k-1), \mathbf{d}(k))$
  $T' = \{e = (s(k-1), a(k), *) \in T\}$
  if $T' = \phi$ **then return** fault: unknown event
  while $isNotEmpty(T')$ **do**
    if $t(k) \in \delta(e = (s(k-1), a(k), s_{new}))$ **then**
      $s(k) := s_{new}$
    else if $t < min(\delta(e))$ **then return** fault: event too early
    else if $t > max(\delta(e))$ **then return** fault: event too late
    $\mu(k) = f_{\Theta_{s(k)}}(y(k-l+1) \ldots y(k-1))$
    $\sigma^2(k) = \frac{\sum_{k=0}^{n-1}\left(y(k) - f_{\Theta_{s(k)}}(y(k-l+1) \ldots y(k-1))\right)^2}{n-1}$
    $F(y(k)) = \frac{1}{2}\left(1 + erf\left(\frac{y(k)-\mu(k)}{\sqrt{2}\sigma(k)}\right)\right)$
    if $F(y(k)) < 1/2 - C/2$ **then return** fault: signal drop
    if $F(y(k)) > 1/2 + C/2$ **then return** fault: signal jump
  end while
end for

Fig. 8. Algorithm 3 Fault detection for observable continuous process variables [20]

Inputs: An array of samples $\{\mathbf{x}_i(k), k = 0, 1, \ldots, n_p - 1\}$ with weights $w_i(k)$.
Outputs: A new array of samples $\mathbf{x}_i(k)$ and their corresponding weights $w_i(k)$.

for each $i \in \{0, \ldots, n_p - 1\}$:
  $z \leftarrow i$
  for $b = 0, \ldots, n_b - 1$:
    $u \sim \mathcal{U}[0,1]$, $j \sim \mathcal{U}\{0, \ldots, n_p - 1\}$
    if $u \leq w_j(k)/w_z(k)$: $z \leftarrow j$
    $a_i \leftarrow z$
for each $i \in \{0, \ldots, n_p - 1\}$: $\mathbf{x}_i(k) \leftarrow \mathbf{x}_{a_i}(k)$, $w_i(k) \leftarrow 1/n_p$

Fig. 9. Algorithm 4 Metropolis Resampling [10]

References

1. Christiansen, L., Fay, A., Opgenoorth, B., Neidig, J.: Improved diagnosis by combining structural and process knowledge. In: Emerging Technologies Factory Automation (ETFA), 2011 IEEE 16th Conference on. pp. 1–8 (2011)
2. Duda, R., Hart, P., Stork, D.: Pattern Classification (2nd Edition). Wiley-Interscience (2000)

3. Frey, C.: Diagnosis and monitoring of complex industrial processes based on self-organizing maps and watershed transformations. In: Computational Intelligence for Measurement Systems and Applications, 2008. CIMSA 2008. 2008 IEEE International Conference on. pp. 87–92 (2008)
4. Gao, Z., Cecati, C., Ding, S.X.: A Survey of Fault Diagnosis and Fault Tolerant Techniques. IEEE Transactions on Industrial Electronics 62(6) (2015)
5. Guo, J., Ji, D., Du, S., Zeng, S., Sun, B.: Fault diagnosis of hybrid systems using particle filter based hybrid estimation algorithm. Chem. Eng. Trans. 33(1), 145–150 (2013)
6. Harris, M.: Optimizing Parallel Reduction in CUDA. NVIDIA Developer Technology
7. Levy, P., Arogeti, S., Wang, D.: An integrated approach to model tracking and diagnosis of hybrid systems. IEEE Transactions on Industrial Electronics 61(4), 1024–1040 (2014)
8. Liu, W., Hwang, I.: Robust estimation and fault detection and isolation algorithms for stochastic linear hybrid systems with unknown fault input. Control Theory Applications, IET 5(12), 1353–1368 (2011)
9. Maier, A.: Online passive learning of timed automata for cyber-physical production systems. In: The 12th IEEE International Conference on Industrial Informatics (INDIN 2014). Porto Alegre, Brazil (2014)
10. Murray, L.M., Lee, A., Jacob, P.E.: Parallel Resampling in the Particle Filter. Accepted at Journal of Computational and Graphical Statistics (2015)
11. Narasimhan, S., Biswas, G.: Model-based diagnosis of hybrid systems. IEEE Transactions on systems, manufacturing and Cybernetics 37, 348–361 (2007)
12. Niggemann, O., Lohweg, V.: On the Diagnosis of Cyber-Physical Production Systems - State-of-the-Art and Research Agenda. In: In:Twenty-Ninth Conference on Artificial Intelligence (AAAI-15) (2015)
13. Ristic, B., Arulampalam, S., Gordon, N.: Beyond the Kalman Filter: Particle Filters for Tracking Applications. Artech House (2004)
14. Shui, A., Chen, W., Zhang, P., Hu, S., Huang, X.: Review of fault diagnosis in control systems. In: Control and Decision Conference, 2009. CCDC '09. Chinese. pp. 5324–5329 (2009)
15. Vodenčarević, A., Kleine Buning, H., Niggemann, O., Maier, A.: Identifying behavior models for process plants. In: Emerging Technologies & Factory Automation (ETFA), 2011 IEEE 16th Conference on. pp. 1–8 (2011)
16. Wang, M., Dearden, R.: Detecting and Learning Unknown Fault States in Hybrid Diagnosis. In: Proceedings of the 20th International Workshop on Principles of Diagnosis, DX09. pp. 19–26. Stockholm, Sweden (2009)
17. Windmann, S., Jiao, S., Niggemann, O., Borcherding, H.: A Stochastic Method for the Detection of Anomalous Energy Consumption in Hybrid Industrial Systems. In: INDIN (2013)
18. Windmann, S., Niggemann, O.: Automatic model separation and application to diagnosis in industrial automation systems. In: IEEE International Conference on Industrial Technology (ICIT 2015) (2015)
19. Windmann, S., Niggemann, O.: Automatic model separation and application to diagnosis in industrial automation systems. In: IEEE International Conference on Industrial Technology (ICIT 2015) (2015)
20. Windmann, S., Niggemann, O.: Efficient Fault Detection for Industrial Automation Processes with Observable Process Variables. In: IEEE International Conference on Industrial Informatics (INDIN 2015) (2015)

21. Windmann, S., Niggemann, O.: A GPU-Based Method for Robust and Efficient Fault Detection in Industrial Automation Processes. In: IEEE International Conference on Industrial Informatics (INDIN 2016) (2016)
22. Yu, M., Wang, D., Luo, M.: Model-based prognosis for hybrid systems with mode-dependent degradation behaviors. IEEE Transactions on Industrial Electronics 61(1), 546–554 (2014)
23. Zhao, F., Koutsoukos, X.D., Haussecker, H.W., Reich, J., Cheung, P.: Monitoring and fault diagnosis of hybrid systems. IEEE Transactions on Systems, Man, and Cybernetics, Part B 35(6), 1225–1240 (2005)

Validation of similarity measures for industrial alarm flood analysis*

Marta Fullen[1] and Peter Schüller[2] and Oliver Niggemann[1,3]

[1] Fraunhofer IOSB-INA Institutsteil für industrielle Automation, Lemgo, Germany
[2] Technische Universität Wien, Institut für Logic and Computation, Vienna, Austria
[3] Institute Industrial IT, Lemgo, Germany

Abstract. The aim of industrial alarm flood analysis is to assist plant operators who face large amounts of alarms, referred to as alarm floods, in their daily work. Many methods used to this end involve some sort of a similarity measure to detect similar alarm sequences. However, multiple similarity measures exist and it is not clear which one is best suited for alarm analysis. In this paper, we perform an analysis of the behaviour of the similarity measures and attempt to validate the results in a semi-formalised way. To do that, we employ synthetically generated floods, based on assumption that synthetic floods that are generated as 'similar' to the original floods should receive similarity scores close to the original floods. Consequently, synthetic floods generated as 'not-similar' to the original floods are expected to receive different similarity scores. Validation of similarity measures is performed by comparing the result of clustering the original and synthetic alarm floods. This comparison is performed with standard clustering validation measures and application-specific measures.

1 Introduction

The phenomenon of alarm flooding is a recurring problem in industrial plant operation [19]. It occurs when the frequency of alarm annunciations is so high that it exceeds the operators capability of understanding the situation. This creates a dangerous situation where the operator might overlook critical alarms that could lead to significant downtime, irreversible damage or even loss of life [10].

The main reason for alarm flooding is imperfect alarm system design. Figure 1 presents a typical alarm generation system consisting of several modules. The alarms are triggered based on sensor values, thresholds or more complex rules. Basic signal and alarm filtering can be used to remove alarms that are known to be noise before they are displayed to the operator. Operators themselves have the opportunity to shelve alarms that they consider irrelevant or redundant. However, the real potential lies in the "contextual preclassification" block, where expert knowledge can be combined with intelligent computational methods to assist the operator.

* This paper is an extended version of [7] and involves a more detailed analysis of the behaviour of similarity measures.

O. Niggemann and P. Schüller (Eds.), *IMPROVE – Innovative Modelling Approaches for Production Systems to Raise Validatable Efficiency*, Technologien für die intelligente Automation 8, https://doi.org/10.1007/978-3-662-57805-6_6

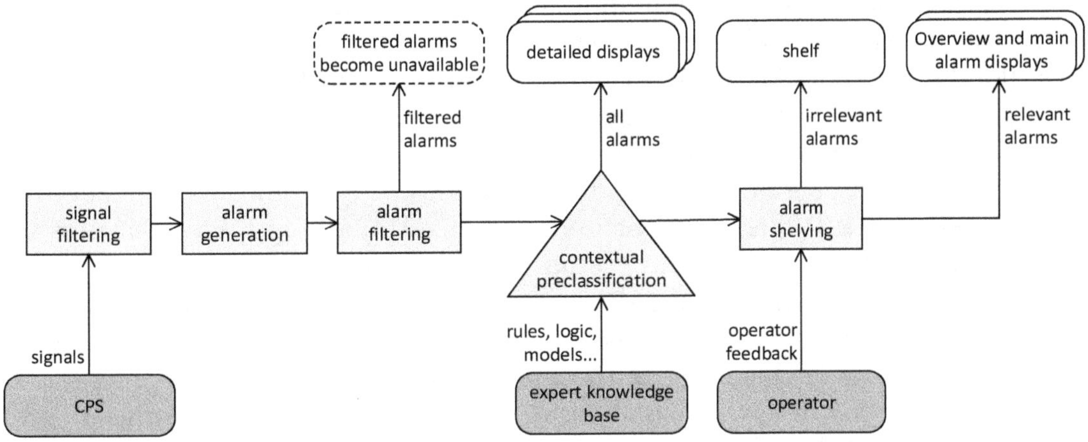

Fig. 1: Typical alarm filtering and shelving system extended by contextual preclassification step (from: [16]).

Diagnosis of a failure of an industrial plant is a non-trivial task that requires extensive knowledge of causalities between symptoms produced by the system [15]. Model-based methods often require explicit representation of expert knowledge about the system and about possible failures, which is rarely available to the extent that is necessary for the method to provide reliable results. In these cases shallow data-driven approaches are more applicable. Data-driven approaches are based purely on the data obtained from the system, possibly with rudimentary expert knowledge inserted when available to improve the results. Data-driven methods directly analyse, manage and reduce the alarm annunciation and therefore flooding [2], without a semantic representation of the system. Multiple approaches exist to this end, drawing from the data mining fields such as sequence identification and pattern recognition [18, 4], correlation analysis [21] or visualisation [13]. Many of these approaches utilise flood similarity measure of some kind, e.g., [20, 3].

Alarm flood detection and clustering is a data-driven approach to handle alarm flooding. An operator assistance system can detect a newly annunciated alarm flood, compare it to the previously seen floods and identify the most similar cluster. If the history of flood of the plant has annotations, such as a log of repairs done to remedy the reason of an original flood, the system can make a suggestion to the operator regarding the fault diagnosis and repair procedure.

A major challenge in creating such a system is determining how similarity between alarm floods should be defined. A multitude of similarity measures (and, analogously, distance measures) exists in the field of data mining and clustering [5]. Another challenge is, that real industrial alarm data is difficult to work with, e.g. because of a high volume of alarms, or because of poor alarm system design. While a certain similarity measure might work for a certain application, there is no guarantee it will work in other application scenarios, as there is not systematic method for quantifying the usefulness of a given similarity measure in the industrial setting.

We here propose a semi-formal approach to answering that issue by defining an experimental design to validate the behaviour of a distance measure in regard to alarm flood clustering. Analysis of the behaviour of the distance measure can then help choose the most suitable distance measure. We also reproduce and extend the alarm flood detection and clustering approach by Ahmed et al. [1] with additional similarity measures based on Term frequency-inverse document frequency (TF-IDF) representation [12] and Levenshtein distance [14], apply these measures to a large real industrial alarm log and evaluate them using our validation method.

Our study shows that the flood distance measure in [1] behaves significantly different from the other analysed distance measures; in particular, the other measures produce a more stable clustering in the presence of noise in the data.

Methodology for detecting floods and computing distance measures is given in Section 2, as well as our proposed approach for validating the behaviour of clustering results using alarm floods synthetically generated from real alarm flood data. Section 4 presents and discusses results of empirical validation on a real industrial dataset. We summarize the results and conclude in Section 5.

2 Clustering methodology

The approach for alarm flood similarity measure analysis is illustrated in Figure 2. Alarm flood similarity measure analysis consists of seven steps divided into two stages: (1) clustering performed once on alarm floods detected in the alarm log and (2) repeatable generation of synthetic alarm floods and clustering. Analysis is preceded by the acquisition of alarm log A from the Cyber—Physical Production System (CPPS) performed by component C1, as described in Section 2.1. Component C2 is responsible for detecting the alarm floods which as described in Section 2.2. Stage 1 clustering (component C4) is performed only once on the set of alarm floods detected in the alarm log F^O and yields a clustering solution S^O. Clustering methodology is described in Section 2.3. Stage 2 is performed multiple times and begins with synthetic flood generation (component C3) which yields a set of synthetic floods F^S. Synthetic floods are clustered alone by component C6 which yields solution S^S, as well as merged with the original floods forming F^{OS} and clustered together by component C5, yielding solution S^{OS}. The procedure for generating synthetic floods is described in Section 3.1. The process is repeated an arbitrary number of times and each round is evaluated according to the metrics described in Section 3.

2.1 Alarm log acquisition

Alarms triggered within a CPPS are processed by an alarm logging system (component C1) within its data acquisition and management system. Alarms are displayed to the operator and saved into a historical database referred to as a historical alarm log A. Recorded alarms are characterised by, at the very least, alarm ID, alarm trigger time and alarm acknowledgement time. Alarm log can also contain additional information such as a description, location and other details of the alarm.

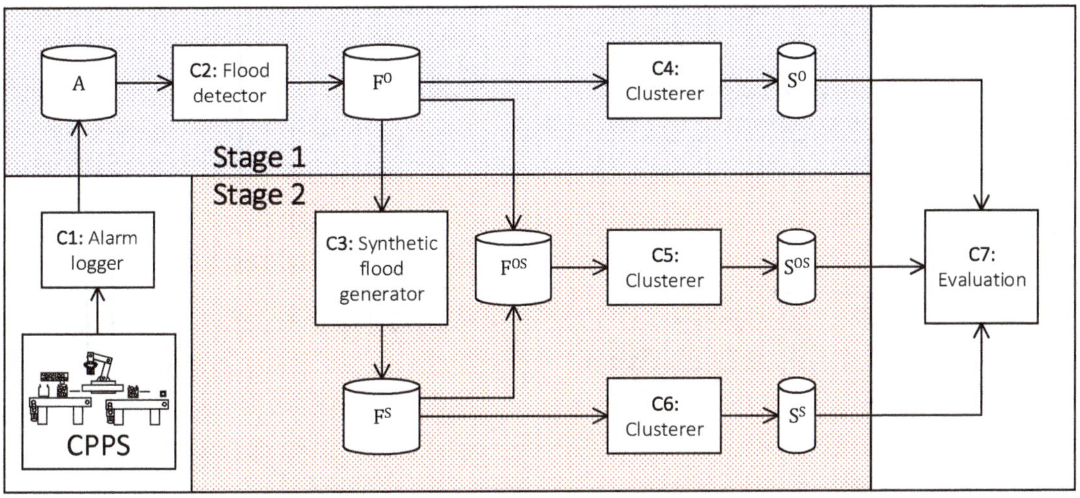

Fig. 2: Approach for validating alarm flood similarity measures.

2.2 Flood detection and preprocessing

First step of the analysis is performed once for a historical alarm log.

Flood detector (component C2) detects the floods based on the alarm flood definition. According to the industry standard for Management of Alarm Systems for Process Industries, an alarm flood begins when the alarm annunciation rate exceeds 10 alarms per 10 minutes and ends when the alarm annunciation rate drops under 5 alarms per 10 minutes [9], see Figure 3c.

However, the specific methodology for flood detection is up to interpretation.

Firstly, the alarm logs should be preprocessed to account for the lingering and chattering alarms (see Figure 3a). Ambiguity arises in the case of lingering alarms that are active for a long time before the flood is considered to begin. Since they might be relevant to the root cause of the flood, we include such alarms in the flood if they were triggered no longer than a lingering threshold t_l before the flood start (see Figure 3b). Moreover, chattering alarms falsely increase the number of alarms within a time period so they are merged before flood detection.

Redundant alarms convey the same information, see Figure 3a, while unnecessarily cluttering the operator display. As a most simple example, two alarms triggered based on the same condition are redundant. In more complex situations, a redundant alarm is triggered based on a condition which is directly caused by an event that triggers another alarm, and is not influenced by any other factors. Redundant alarms are more difficult to deal with, requiring an analysis and refinement of the alarm system or using an advanced reasoning to detect causalities between alarms.

Floods are detected using a sliding window and the outcome is a set of original floods F^O.

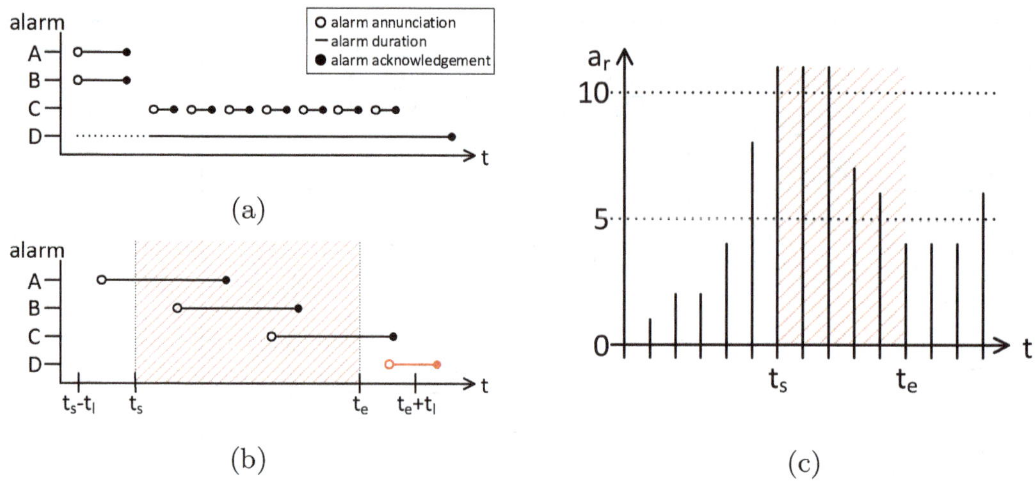

Fig. 3: (a) Undesirable alarms caused by imperfections in alarm system design: A, B - redundant alarms, C - chattering alarm, D - lingering alarm. (b) Flood detection from a more detailed point of view, where a dilemma arises which alarms exactly should be included. Alarms A, B, C are included in the flood, while alarm D is not. t_s - alarm flood start, t_e - alarm flood end, t_l - lingering alarm inclusion margin. (c) Flood detection: a_r - alarm rate (alarms active in 10 minute period), t - 10-minute time periods, t_s - alarm flood start, t_e - alarm flood end.

2.3 Alarm flood clustering

Original flood set F^O is clustered to obtain the original clustering solution S^O (see Figure 2 component C4). Density-based spatial clustering of applications with noise (DBSCAN) [6] is a clustering algorithm that intrinsically deduces the most-fitting number of clusters. It is based on the concept of density, where points within a specified distance threshold ϵ to each other are considered to belong to a dense area—a cluster. Points that are distant from dense areas are considered outliers and are gathered in a separate group. Preliminary experiments showed, that it is a disadvantage to predefine the number of clusters, because the number of natural clusters in the data is expected to change with different distance measures. Therefore, to perform an unbiased comparison of distance measures, we use the DBSCAN algorithm which is not biased to a fixed number of clusters and exclude the outliers in our analysis.

Choice of a distance measure is the second, after the clustering algorithm itself, most critical aspect when performing clustering, and the focus of this paper. Each distance measure requires a specific data representation and four distance measures are analysed: Jaccard distance [11] on a bag of words representation, distance based on the frequency of consecutive alarms [1], Euclidean distance on TF-IDF representation [12] and Levenshtein distance [14]. Chosen measures are focused either on the appearance of alarms (Jaccard distance and TF-IDF representation) or on the order of alarms (frequency of consecutive alarms and Levenshtein distance); in the latter, the absolute time distance is not considered.

Jaccard distance (J) Jaccard distance is the ratio between alarm types occurring only in one of the two floods, and the number of alarm types in both floods. Each flood f_i is represented as a binary vector $f_i = (a_1, a_2, \ldots, a_m)$, where m is the number of unique alarm identifiers in the complete alarm log and a_j is a binary value representing whether an alarm appeared in the flood, regardless of its count. The Jaccard distance between floods f_i and f_j is given by

$$J_{ij} = \frac{|f_i \text{ xor } f_j|}{|f_i \text{ or } f_j|}, \tag{1}$$

where $|x|$ returns the number of true values in vector x. Alarm types that are absent from both floods are irrelevant for Jaccard distance. This measure was used as preprocessing in [1].

Frequency of consecutive alarms (F) This measure was proposed in [1] based on a simplification of first-order Markov chains. Each flood is represented as a matrix of counts of each pair of alarms appearing consecutively,

$$P = \begin{bmatrix} f_{11} & f_{12} & \cdots & f_{1m} \\ \vdots & \vdots & \ddots & \vdots \\ f_{m1} & f_{m2} & \cdots & f_{mm} \end{bmatrix}, \tag{2}$$

where f_{ij} is the frequency of alarm a_j being annunciated directly after alarm a_i in a given alarm flood. Then, the distance between two floods can be calculated as a distance between their P matrices, e.g. using Frobenius distance.

Term frequency-inverse document frequency (T) TF-IDF is a measure often used in natural language processing to weight terms in a document according to how frequent and discriminative they are with respect to a document collection. We apply TF-IDF to weight alarms in alarm floods with respect to the collection of all floods. TF-IDF is calculated for each alarm a and flood f as

$$\text{tf-idf}(a, f) = \text{tf}(a, f) * \text{idf}(a). \tag{3}$$

Term frequency is calculated as

$$\text{tf}(a, f) = \frac{f_{a,f}}{|f|}, \tag{4}$$

where $f_{a,f}$ is the number of annunciations of alarm a in flood f and $|f|$ is the total number of alarm annunciations in f.

Inverse document frequency is calculated as

$$\text{idf}(a) = \log_e \frac{|F|}{|\{f \in F \mid a \in f\}|}, \tag{5}$$

where $|F|$ is the total number of floods. TF-IDF score is calculated for every alarm type and every flood in the log and yields a flood representation in the form of a vector of length m, the total number of unique alarm signatures.

Two floods f_i and f_j can then be compared using a distance measure such as Euclidean distance between two vectors:

$$d(f_i, f_j) = \sqrt{\sum_{k=1}^{m} (\text{tf-idf}(a_k, f_i) - \text{tf-idf}(a_k, f_j))^2}. \tag{6}$$

Levenshtein distance (L) This metric counts the amount of "edits" that are needed to transform one sequence into another one, where an edit is a symbol insertion, symbol deletion, or a symbol substitution. To apply this metric, alarm floods are represented as sequences of symbols, which in turn represent unique alarm types. The Levenshtein distance $d(|f_i|, |f_j|)$ between floods f_i and f_j is calculated recursively, where the distance for the first x and y symbols of f_i and f_j, respectively, is calculated as follows:

$$d(x,y) = \begin{cases} \max(x,y) & \text{if } \min(x,y) = 0 \\ \min \begin{cases} d(x-1,y)+1 \\ d(x,y-1)+1 \\ d(x-1,y-1)+\mathbf{1}_{f_i(x) \neq f_j(y)} \end{cases} & \text{otherwise,} \end{cases} \tag{7}$$

where $\mathbf{1}_{condition}$ is the indicator function. The distance score is normalised over the length of floods.

2.4 Distance matrix postprocessing

Distance matrix is optionally postprocessed using a Jaccard similarity measure threshold t. The rationale for postprocessing is that only floods that have a significant number of alarms in common can be assigned a low distance value [1]. To ensure that, distance values for each pair of floods are filtered based on the value of the Jaccard distance for this pair. For any distance measure d, distance value d_{ij} remains unchanged if the Jaccard distance between the corresponding floods d_{ij}^J is lower than the threshold t; otherwise, the distance is replaced with the maximum value $d_{ij} = 1$:

$$\hat{d}_{ij} = \begin{cases} d_{ij} & \text{if } d_{ij}^J < t, \\ 1 & \text{otherwise.} \end{cases} \tag{8}$$

Postprocessed distance matrix $\hat{d}$ yields clustering solution $\hat{S}$.

3 Evaluation methodology

At the core of the proposed approach, the original flood set F^O is concatenated with the synthetic flood set F^S to create a joint set F^{OS}. Procedure to generate synthetic floods is described in section 3.1. The clustering result of the join set, S^{OS} is used to evaluate the behaviour of the distance measures. Clustering (see Figure 2 components 5 and 6) is performed analogously to the original floods, as described in 2.3. We propose to evaluate the clustering solutions in regard to the distance measure behaviour using two measures: cluster membership of synthetic floods m_1 (see section 3.2) and cluster stability, based on adjusted Rand index [17], in five variants: R_1, R_2, R_3, R_4 and R_5 (se section 3.3).

3.1 Synthetic flood generation

Second step of the analysis is focused on generating synthetic alarm floods that can be included in clustering to evaluate the distance measures. Synthetic floods are generated based on the existing set of original floods F^O. Each original flood f^O is used to create one synthetic flood f^S. The original alarm flood is modified in three ways to create the synthetic flood, as illustrated in Figure 4: (i) by addition of randomly chosen alarms, (ii) by removal of randomly chosen alarms and (iii) by transposing randomly chosen pairs of alarms.

Those three modifications correspond to the expected possible variations in an industrial alarm log, which stem e.g. from delays on the bus or data acquisition sampling rate. For simplicity, we always apply an equal amount, at least one, of all three modifications to each flood. The number of modifications is varied throughout the experiments to obtain different synthetic flood sets, ranging from very similar to dissimilar to the original floods. We represent the degree of modification as a percentage of the number of alarms in a flood that has been modified.

This way of modification of floods from real datasets is chosen according to our domain knowledge and experience with floods and their variation in the industrial setting.

3.2 Cluster Membership of Synthetic Floods

The first validation approach is the fraction of synthetic floods that is assigned to the same cluster as their original flood. It is calculated as

$$m_1 = \frac{|f_i^S : c(f_i^S) = c(f_i^O)|}{|F^S|}, \tag{9}$$

where f_i^S is a synthetic flood generated from original flood f_i^O, $c(f)$ is the cluster flood f was assigned to and F^S is the set of all synthetic floods. This measure is calculated for each synthetic flood with respect to its mother flood, without considering (potentially random) similarities to other original floods. Therefore, if a synthetic flood by chance becomes more similar to a different original flood than its mother flood, it will not affect the results.

3.3 Cluster Stability

We can consider the original flood clustering results to be the ground truth, or the "target", as in the supervised machine learning validation methods. This assumption is made only for the purpose of validation of the similarity measure behaviour.

Each synthetic flood has a known mother flood, and is expected to be treated similarly by the clustering algorithm if it was generated with a low degree of modification; i.e., the synthetic flood is expected to have similar distance scores to other floods as its mother flood, and therefore to be clustered alike. Hence, the synthetic floods are given the same target as their mother floods. On the other hand, if the synthetic flood was generated using a high degree of modification it is expected to be treated differently by the clustering algorithm than its original flood.

Results of clustering original and synthetic floods together can be compared to that target solution. Adjusted RAND index is a well-known measure for quantifying partition agreement between clustering solutions, while disregarding the actual cluster number. Adjusted RAND index for two partitions $C_1 = \{c_0, c_1, ..., c_n\}$ and $C_2 = \{c_0, c_1, ..., c_n\}$ of items is calculated as

$$R = \frac{a+b}{a+b+c+d}, \tag{10}$$

where a is the number of pairs of items that are in the same cluster in C_1 and in C_2, b is the number of pairs of items that are in different clusters in C_1 and in C_2, c is the number of pairs of items that are in the same cluster in C_1 but in different clusters in C_2, and d is the number of pairs of items that are in different clusters in C_1 but in the same cluster in C_2.

We specify the following five different variants of cluster stability to perform the analysis. Cluster stability R_1 compares the cluster membership of original floods and synthetic floods when clustered together. Cluster stability R_2 is used to quantify the change in original flood partitioning when clustered with and without the synthetic floods. Cluster stability R_3 quantifies the difference between the expected outcome, which is the result of the original flood clustering, and the obtained solution. Cluster stability R_4 compares the cluster memberships of the original and the synthetic floods when clustered separately. Finally, cluster stability R_5 compares the solutions for the original floods obtained with and without the postprocessing step described in section 2.4.

$$R_1 = R(S^{OS}(0,n), S^{OS}(n+1, n+m)), \tag{11}$$

$$R_2 = R(S^O, S^{OS}(0,n)), \tag{12}$$

$$R_3 = R(S^{OS}, S^O \cdot S^O), \tag{13}$$

$$R_4 = R(S^O, S^S), \tag{14}$$

$$R_5 = R(S, \hat{S}), \tag{15}$$

where $S^O = s_0^O, ..., s_n^O$ is the clustering solution for the set of original floods $F^O = f_0^O, ..., f_n^O$, $S^S = s_0^S, ..., s_m^S$ is the clustering solution for the set of synthetic floods $F^S = f_0^S, ..., f_m^S$, $s^{OS} = s_0^{OS}, ..., s_n^{OS}, s_{n+1}^{OS}, ..., s_{n+m}^{OS}$ is the clustering solution for the joint set of original and synthetic floods $F^{OS} = f_0^{OS}, ..., f_n^{OS}, f_{n+1}^{OS}, ..., f_{n+m}^{OS}$, $S^{OS}(0,n)$ and $S^{OS}(n+1, n+m)$ denote the joint solution subsets corresponding to the original and synthetic floods respectively and $\hat{S}$ denotes a solution obtained from a postprocessed distance matrix.

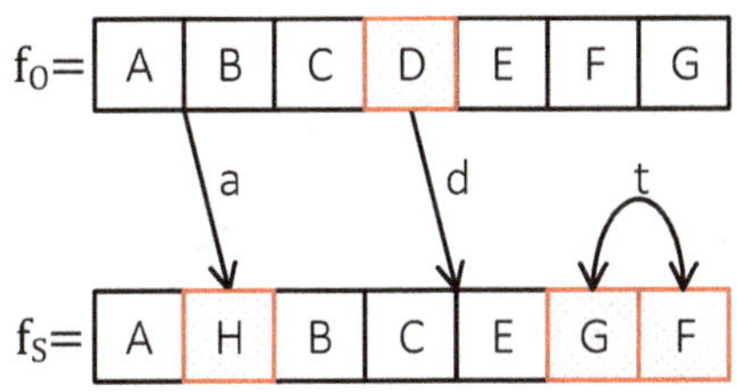

Fig. 4: Creation of a synthetic flood f_s from an original flood f_o composed of alarms A, B, C, D, E, F, G through three edits: a - addition, d - deletion, t - transposition.

Table 1: Specification of performed experiments.

Exp.	Dist. measure	Postproc.
1	Jaccard	No
2	Jaccard	Yes
3	Frequency	No
4	Frequency	Yes
5	TF-IDF	No
6	TF-IDF	Yes
7	Levenshtein	No
8	Levenshtein	Yes

4 Empirical evaluation results

The following results extend those already published in [7]. The experimental data set is a 25-day alarm log from a production plant from the manufacturing industry, consisting of 15 k annunciations of 96 alarm types. The flood detection algorithm has been modified to account for the lingering alarm problem (cf. Section 2.2) and yielded 166 alarm floods with an average length of 61 alarms. We perform similarity measure analysis as described in the methodology section and analyse the behaviour of the distance measure as it changes with adding synthetic floods to the dataset and influences the structure of the clusters. DBSCAN clustering is used in the experiments and since it does not require specification of the number of target clusters, it is possible to observe how do the synthetic floods change the structure of the data, for example by creating new inherent clusters.

4.1 Visualization on a demonstrative set of 25 floods

Fig. 5 visually demonstrates clustering and validation methodology on a reduced dataset of 25 floods. Floods are arranged on X- and Y-axis in the same order, and pixels indicate degree of similarity, where white means highest distance, and strong colour means equality (which occurs mostly on the diagonal).

Row (a) presents original distance matrices of distance measures J, F, T, L. Jaccard distance identified four floods as exactly the same (solid square in the image), while all other distance metrics show that they are in fact not identical: albeit they are composed of the same alarms, they differ in the number and order of their annunciations.

Row (b) shows distance matrices postprocessed according to 2.4. Clearly this filters out many values in the matrices, in particular in the case of TF-IDF many low distance values are reset to highest distance. This property of TF-IDF can be explained, because the principle of TF-IDF is to put more weight on terms that occur more often in one flood and that occur less often in other floods. Therefore, to

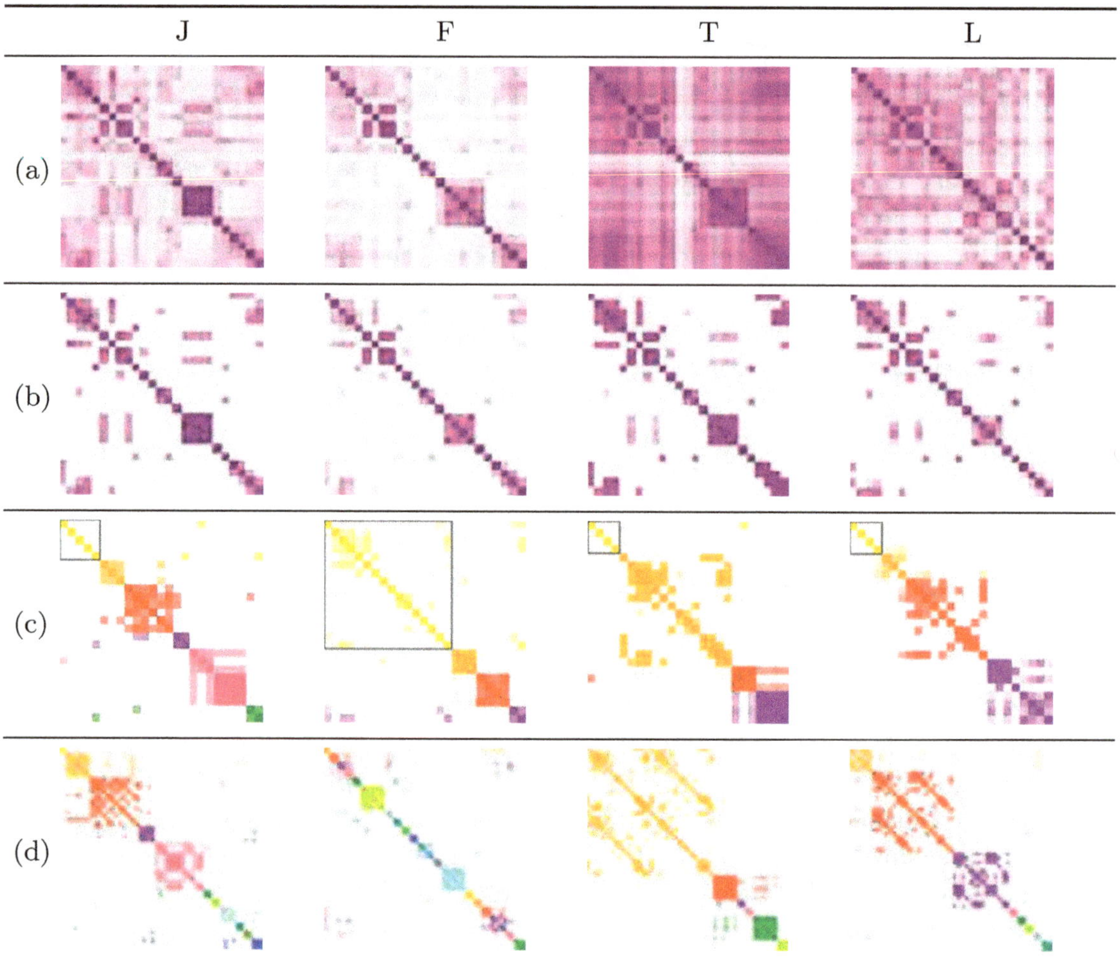

Fig. 5: Distance matrices obtained at every stage of the experiments for measures J, F, T, and L for a reduced dataset of 25 floods. Pixels indicate pairwise similarity between floods which are arranged on X- and Y-axis in the same order: stronger colours show lower distance. Rows show (a) original flood distance matrices; (b) distance matrices postprocessed using Jaccard distance threshold; (c) clustering solutions using distances from (b), coloured according to the cluster number, with outliers in the top left cluster highlighted by a black border; and (d) clustering solutions after adding a 10% modified synthetic flood for each flood.

obtain high distance, a pair of floods needs to contain a distinct set of alarms that has low frequency in other floods. In our dataset this rarely happens, concretely it mainly happens in short floods, where the term frequency of rare alarms contributes a large value to the distance measure.

Row (c) presents the DBSCAN clustering results on distance matrices of row (b), where floods have been rearranged and coloured according to the cluster number. The top left cluster represents outliers: their distance to other floods was under the ϵ threshold and therefore they were not assigned to any cluster.

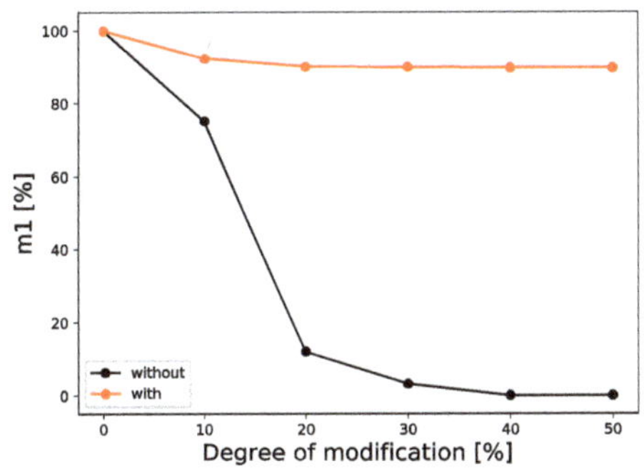

Fig. 6: Measure m_1 - cluster membership of synthetic floods calculated with and without outliers for frequency of consecutive alarms distance measure.

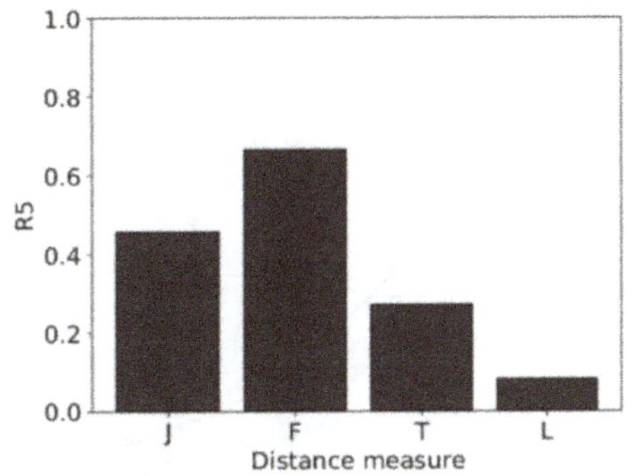

Fig. 7: Measure R_5 - cluster stability between clustering solutions with and without postprocessing calculated for the original flood set.

Row (d) presents clustering results after introducing synthetic floods with 10% modification. Synthetic floods cause changes in the cluster structure, although the structure of original clusters is mostly retained. As in this case, synthetic floods are quite similar to their respective mother floods, many of the outlier floods are clustered together with their synthetic counterpart and form two-flood clusters.

4.2 Clustering with synthetic floods on the full dataset

Further experiments were performed on the whole alarm flood dataset. Each distance measure is evaluated with and without the postprocessing step, as listed in table 1. For validation experiments we generate experimental sets of synthetic floods with a degree of modification (the amount of edits) ranging between none and 50%. Each experiment is repeated 10 times and the measure values are averaged.

Figure 6 shows cluster membership of synthetic floods measure m_1 calculated for the frequency of consecutive alarms distance measure under two conditions: including and excluding the outliers. Since the amount of outliers is high in a real industrial dataset, including them in the analysis distorts the results. The number of classifications that are correct from the perspective of m_1 measure (so number of synthetic floods assigned to the same cluster as the corresponding original flood) is very high because the outliers constitute such a large cluster and the synthetic floods based on the outliers also are assigned to the outlier cluster. Therefore, in further analysis we focus on the floods that were not outliers.

Figure 7 presents the cluster stability between solutions obtained from unchanged distance matrices and postprocessed distance matrices for the original flood set. Low values of R_5 indicate that the solutions obtained with and without postprocessing are not consistent. Results imply that postprocessing heavily influences the results.

Figures 8a to 8d present the evaluation measures m_1, R_1, R_2, R_3 and R_4 for experiments with varying degrees of modification used to generate the synthetic

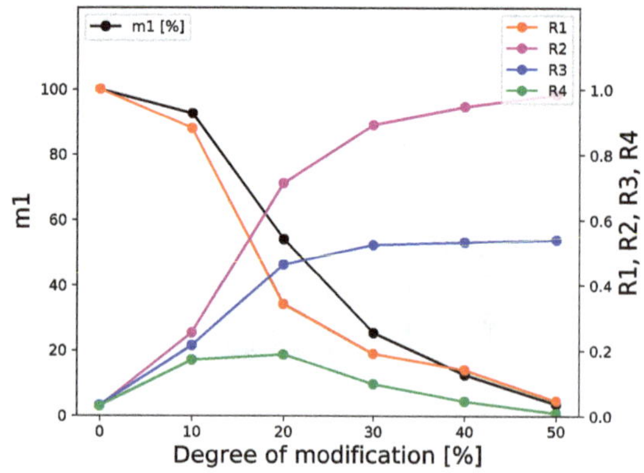

(a) Experiment 1 - Jaccard distance measure without postprocessing.

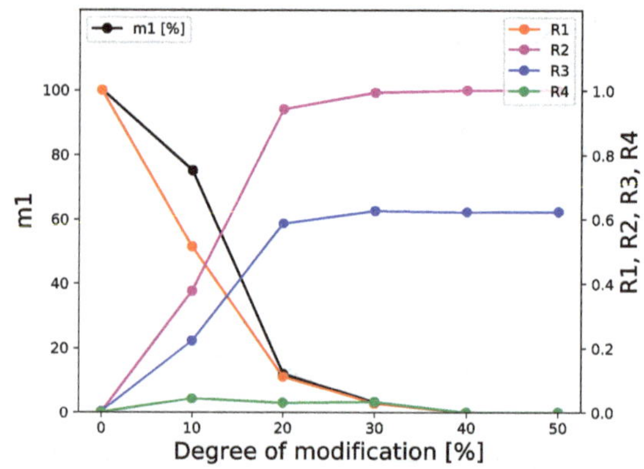

(b) Experiment 3 - Frequency of consecutive alarms distance measure without postprocessing.

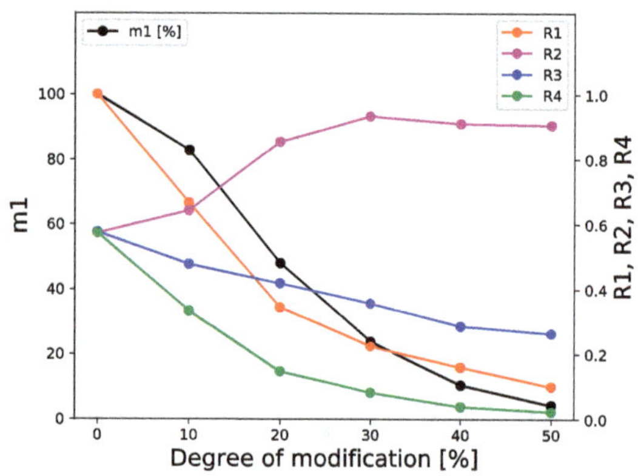

(c) Experiment 5 - TF-IDF distance measure without postprocessing.

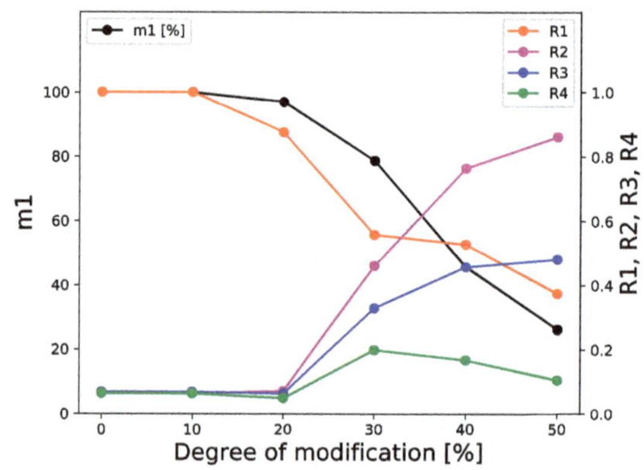

(d) Experiment 7- Levenshtein distance measure without postprocessing.

Fig. 8: Experimental results.

flood sets. Intuitively, identical floods will be clustered together, and indeed for every distance measure the m_1 and R_1 measures for synthetic floods with no modifications are maximal. As the degree of modification is raised, m_1 and R_1 both decrease. The decrease is most abrupt for frequency of consecutive alarms distance measure, which implies that this measure is most sensitive to even small variation in the data.

On the other hand, measures R_2 and R_3 increase with the degree of modification. As the synthetic floods become more and more different from the original floods, the clustering solution of the original floods resembles more the solution obtained in step 1 in the process. That means the synthetic floods no longer sufficiently resemble the original floods. Measure R_3 does not reach the upper limit of 1.0 like measure R_2 does. That is due to the parameter of the clustering algorithm which specifies the minimum number of samples to form a valid cluster. The set of original floods is smaller and therefore the minimum number of samples to form

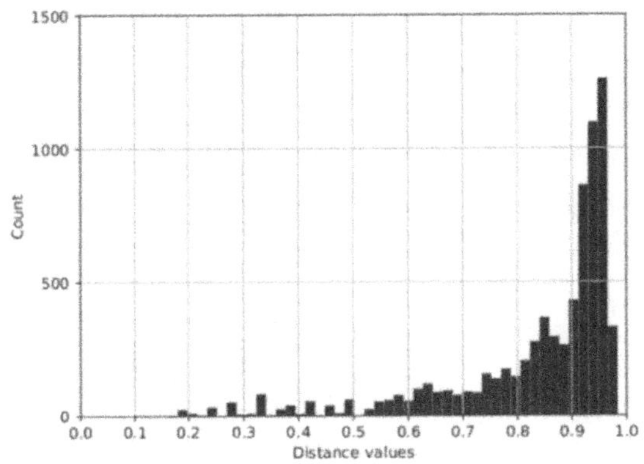

(a) Experiment 1

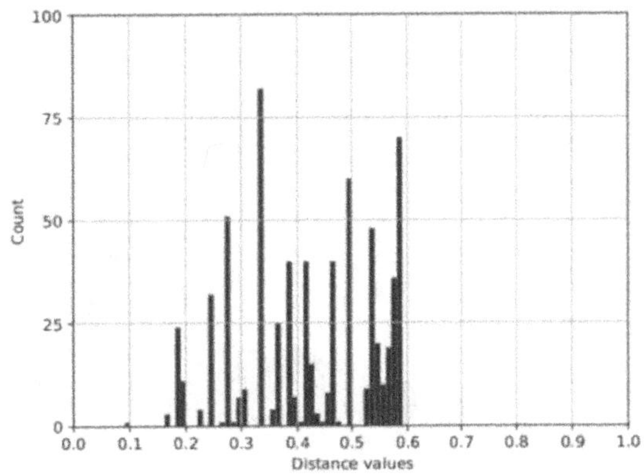

(b) Experiment 2

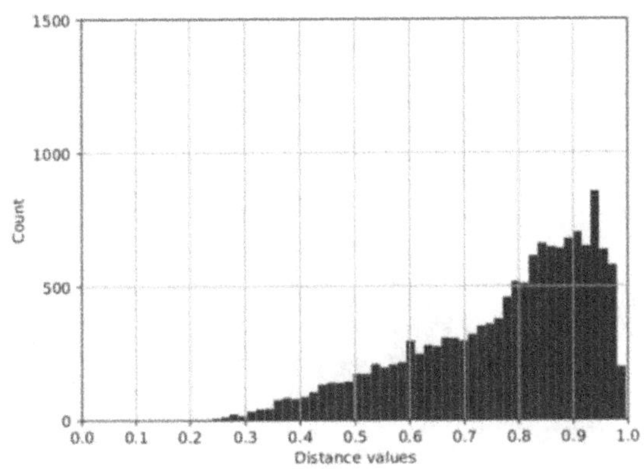

(c) Experiment 3.

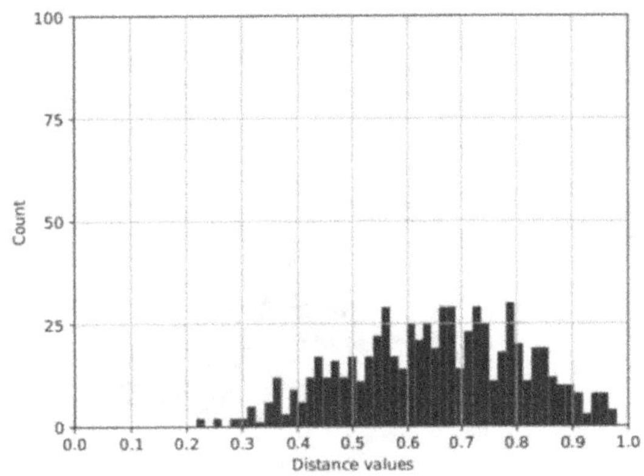

(d) Experiment 4.

Fig. 9: Comparison of distance value histograms before (left) and after (right) post-processing for distance measure: a,b - Jaccard, c,d - Frequency of consecutive alarms. Postprocessed histograms (right) ignore the distance values of 1.

a cluster might not be reached for some subsets of floods, while after adding the synthetic floods, the minimum is reached and clusters are formed.

The effect of postprocessing is analysed further by comparison of histograms of distance matrices before and after postprocessing (Figures 9 and 10). During postprocessing we replace with 1 all the values that have the Jaccard distance value higher than the threshold $t = 0.6$. This means above a certain distance we assume it is maximum distance. The resulting histograms are dominated by these "1" values so we do not show them. We apply this postprocessing to the other distance matrices as well. We always use Jaccard distance as the criterion whether to reset a distance value to 1 (this is the methodology described in [1]) and therefore histograms for the other distance measures show values above 0.6.

In the case of Jaccard distance measure (Figure 9b), postprocessing simply removes all the values in range of $(0.6, 1.0)$. While the postprocessed Levenshtein distance matrix (Figure 10d) is consistent with the Jaccard measure (in the sense that very few distance values remain in the range of $(0.6, 1.0)$), the two other

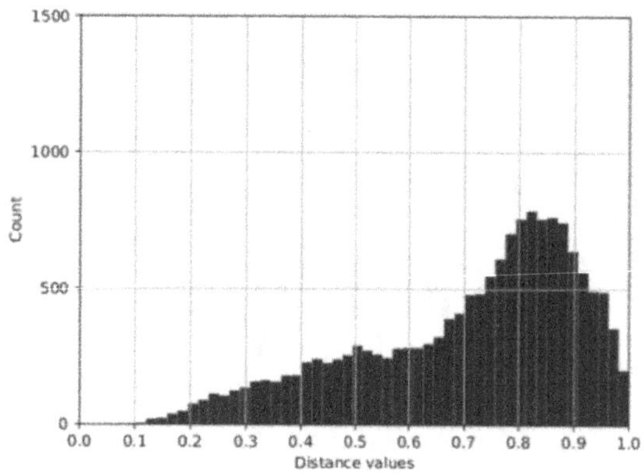

(a) Experiment 5.

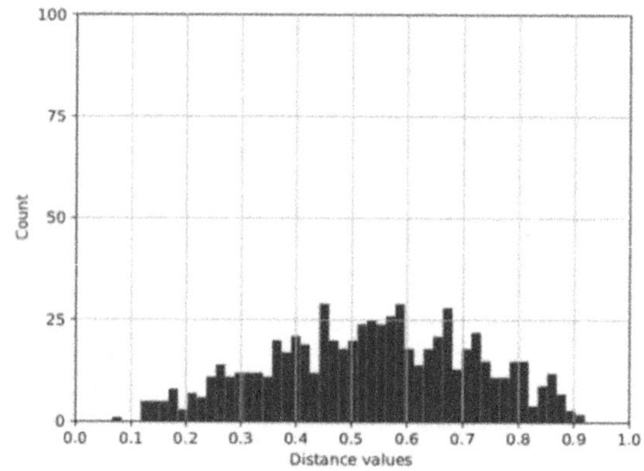

(b) Experiment 6.

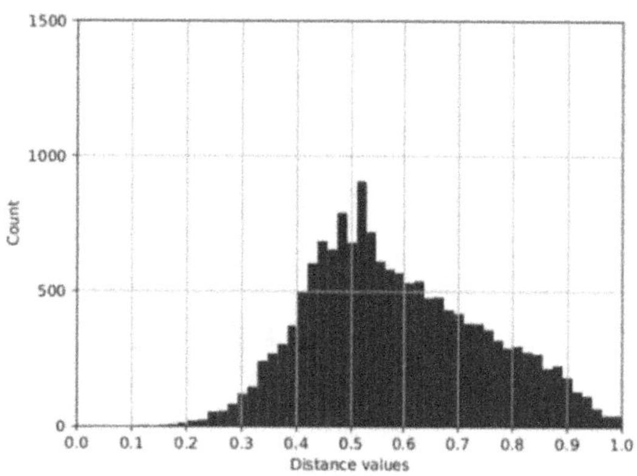

(c) Experiment 7.

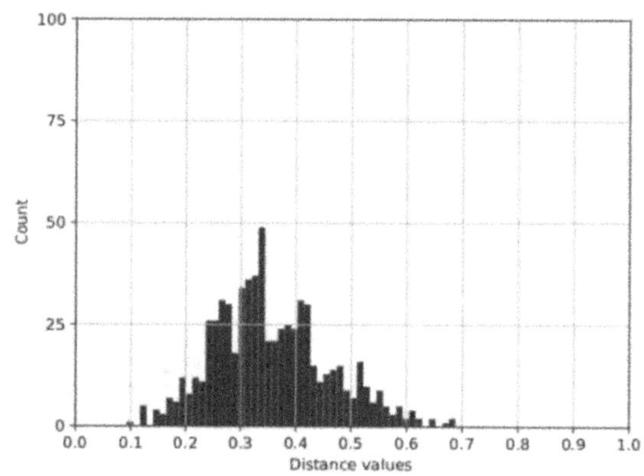

(d) Experiment 8.

Fig. 10: Comparison of distance value histograms before (left) and after (right) postprocessing for each distance measure: a,b - TF—IDF, c,d - Levenshtein. Postprocessed histograms (right) ignore the distance values of 1.

measures show that alarm floods that have many IDs in common may in fact be quite distant in the terms of (i) how frequently alarms appear in order (Figure 9d) and (ii) the most discriminative alarms (Figure 10b).

5 Conclusion

In this paper, we continue the analysis of similarity measures that can be used in alarm flood clustering [7].

The paper presents a methodology for validating similarity measures to help choose a measure which is best suited for similarity-based approaches used in alarm flood analysis. An example of a similarity-based approach is case-based reasoning, where a new alarm flood is compared to a database of known cases to suggest a course of action to the operator [8]. Choice of a similarity measure is a difficult problem, which is normally solved arbitrarily by an expert. Synthetically generated alarm floods create a controlled environment for observing the behaviour and sensitivity of the similarity measure to expected variations in the data, which is not possible using industrial data sets which are generated under unknown conditions and limited in size. The validation process has been shown using a real industrial dataset.

We have previously shown that the measure introduced in [1] produces very different results than our newly introduced measures and results suggest that the measure of [1] is less favourable. TF-IDF representation with Euclidean distance has shown the most balanced results.

Since the alarm logging systems are flawed we can assume the alarm log data to exhibit some degree of variability. For example, there may be delays on the bus or sampling may be too slow so that many alarms are logged with the same timestamp but not necessarily the correct order. Two of the analysed measures (frequency of consecutive alarms and Levenshtein distance) rely heavily on the order of alarms in the data. Results show that frequency of consecutive alarms distance measure is most sensitive to variations in data, while Levenshtein distance is not because it allows transpositions of adjacent alarms within a flood.

Results show that postprocessing heavily influences the results. It has been shown that floods containing many of the same alarm IDs may in fact be very distant when considering other characteristics, such as the order of the alarms or the discriminating value of alarm IDs.

Moreover, DBSCAN clustering appears to produce more meaningful results because of its adaptive choice of the number of clusters.

In the future work, using an annotated dataset (e.g. data generated in a controlled simulation environment) could help further analyse these effect and establish whether postprocessing is viable. Furthermore, additional similarity measures can be analysed, including measures that take absolute time distance into account.

Acknowledgement

This project has received funding from the European Union's Horizon 2020 research and innovation programme under grant agreement No. 678867.

References

1. Ahmed, K., Izadi, I., Chen, T., Joe, D., Burton, T.: Similarity analysis of industrial alarm flood data. In: IEEE Transactions on Automation Science and Engineering (Apr 2013)
2. Bergquist, T., Ahnlund, J., Larsson, J.E.: Alarm reduction in industrial process control. In: Proc. IEEE Conference on Emerging Technologies and Factory Automation. vol. 2, pp. 58–65 (2003)
3. Charbonnier, S., Bouchair, N., Gayet, P.: A weighted dissimilarity index to isolate faults during alarm floods. Control Engineering Practice 45, 110–122 (2015)
4. Charbonnier, S., Bouchair, N., Gayet, P.: Fault template extraction to assist operators during industrial alarm floods. Engineering Applications of Artificial Intelligence 50, 32–44 (2016)

5. Deza, M.M., Deza, E.: Encyclopedia of Distances. Springer (2009)
6. Ester, M., Kriegel, H.P., Sander, J., Xu, X.: A density-based algorithm for discovering clusters in large spatial databases with noise. In: KDD. pp. 226–231. AAAI Press (1996)
7. Fullen, M., Schüller, P., Niggemann, O.: Defining and validating similarity measures for industrial alarm flood analysis. In: IEEE 15th International Conference on Industrial Informatics (INDIN) (2017)
8. Fullen, M., Schüller, P., Niggemann, O.: Semi-supervised case-based reasoning approach to alarm flood analysis. In: Machine Learning for Cyber Physical Systems (ML4CPS) (2017)
9. Instrumentation, Systems, and Automation Society: ANSI/ISA-18.2-2009: Management of Alarm Systems for the Process Industries (2009)
10. Izadi, I., Shah, S.L., Shook, D.S., Chen, T.: An introduction to alarm analysis and design. In: IFAC SAFEPROCESS. pp. 645–650 (2009)
11. Jaccard, P.: Distribution de la flore alpine dans le bassin des Dranses et dans quelques régions voisines. Bulletin de la Société Vaudoise des Sciences Naturelles 37, 241–272 (1901)
12. Jones, K.S.: A statistical interpretation of term specificity and its application in retrieval. Journal of Documentation 28, 11–21 (1972)
13. Laberge, J.C., Bullemer, P., Tolsma, M., Reising, D.V.C.: Addressing alarm flood situations in the process industries through alarm summary display design and alarm response strategy. Intl. J. of Industrial Ergonomics 44(3), 395–406 (2014)
14. Levenshtein, V.I.: Binary codes capable of correcting deletions, insertions and reversals. Soviet Physics Doklady 10(8), 707–710 (1966)
15. Niggemann, O., Lohweg, V.: On the diagnosis of cyber-physical production systems: State-of-the-art and research agenda. In: Proc. AAAI. pp. 4119–4126. AAAI Press (2015)
16. Norwegian Petroleum Directorate: YA-711 Principles for alarm system design (2001)
17. Rand, W.M.: Objective criteria for the evaluation of clustering methods. Journal of the American Statistical association 66(336), 846–850 (1971)
18. Vogel-Heuser, B., Schütz, D., Folmer, J.: Criteria-based alarm flood pattern recognition using historical data from automated production systems (aps). Mechatronics 31, 89–100 (2015)
19. Wang, J., Yang, F., Chen, T., Shah, S.L.: An overview of industrial alarm systems: Main causes for alarm overloading, research status, and open problems. IEEE Transactions on Automation Science and Engineering 13(2), 1045–1061 (2016)
20. Wang, J., Li, H., Huang, J., Su, C.: A data similarity based analysis to consequential alarms of industrial processes. Journal of Loss Prevention in the Process Industries 35, 29–34 (2015)
21. Yang, F., Shah, S., Xiao, D., Chen, T.: Improved correlation analysis and visualization of industrial alarm data. ISA Transactions 51(4), 499–506 (2012)

Concept for Alarm Flood Reduction with Bayesian Networks by Identifying the Root Cause

Paul Wunderlich and Oliver Niggemann

Ostwestfalen-Lippe University of Applied Sciences, Institute Industrial IT, Langenbruch 6, 32657 Lemgo, Germany
{paul.wunderlich,oliver.niggemann}@hs-owl.de
http://www.init-owl.de

Abstract. In view of the increasing amount of information in the form of alarms, messages or also acoustic signals, the operators of systems are exposed to more workload and stress than ever before. We develop a concept for the reduction of alarm floods in industrial plants, in order to prevent the operators from being overwhelmed by this flood of information. The concept is based on two phases. On the one hand, a learning phase in which a causal model is learned and on the other hand an operating phase in which, with the help of the causal model, the root cause of the alarm sequence is diagnosed. For the causal model, a Bayesian network is used which maps the interrelations between the alarms. Based on this causal model the root cause of an alarm flood can be determined using inference. This not only helps the operator at work, but also increases the safety and speed of the repair. Additionally it saves money and reduces outage time. We implement, describe and evaluate the approach using a demonstrator of a manufacturing plant in the SmartFactoryOWL.

1 Introduction

The next industrial revolution (digital transformation) not only affects society but also the working world. In the future, a few plant operators will have to operate highly automated and complex plants. Especially, the vision of complete networking of all components and parts of a plant (IoT) leads to overloaded operators due to the enormous load of information. This is particularly critical in the area of the alarm management of plants and machines, since none, too late or incorrect intervention can result in high material damage or even personal injury. As a result of the increasing automation, additional sensors are increasingly installed because of their allegedly very good ratio of security per cost. However, this results in an enormous number of warnings and alarms, which overexert the operator [15]. Such situations are called alarm floods. As an effect of this, the operator is only able to acknowledge most of the alarms, but cannot process the information that they provide. This may cause dramatic effects especially at high hazard facilities like in the process industry and reduces the overall value of an alarm system (see Example: Refinery Explosion).

The Author(s) 2018
O. Niggemann and P. Schüller (Eds.), *IMPROVE – Innovative Modelling Approaches for Production Systems to Raise Validatable Efficiency*, Technologien für die intelligente Automation 8, https://doi.org/10.1007/978-3-662-57805-6_7

An alarm flood corresponds to a time interval in which the number of alarms is higher than the reactivity of an operator. There exist different reasons for an alarm flood. Most common is a badly designed alarm management system. Based on [21, 24, 28] typical distinctive features of an inefficiently designed alarm management are: missing alarm philosophy, irrelevant alarms, chattering or nuisance alarms, incorrectly configured alarm variables, alarm design isolated from related variables, permanent alarms in normal state, alarms and warnings at the same time, missing option to remove remedied alarms and too many priorities.

Therefore an effective alarm management for factories is of huge interest and current research topic. Regardless of very good alarm management, the weaknesses of humans are that they are not able to perform 100 percent consistently well all the time. For example, if operators are tired, sick, distracted or stressed, their performance might be degraded. Therefore, even with well managed alarm systems, it is recommended to limit the credit given to alarms [13]. However, despite an effective alarm system design and configuration efforts, the occurrence of alarm flooding cannot be eliminated completely [5]. The spread of alarms has become one of the biggest problems for plant operators in modern times. Therefore, it is essential to protect the operator from unnecessary information and to help him focus on his task. One way to accomplish it, is to reduce the amount of alarms in an alarm flood. In particular, sequences of alarms, which are connected in a causal context, are very suitable because they cannot be excluded despite a good alarm management. This alarm sequence is reduced so that only the alarm which points to the cause is displayed. For this purpose the root cause of the alarm flood must be identified. The root cause is an initiating cause of a causal chain which leads to an alarm flood. To achieve this, we developed a concept based on a causal model of the interrelations of the alarms. This is discussed in more detail in Section 4. We use Bayesian networks, which where developed by J. Pearl et al. [22] as a causal model. With the causal model we determine the potential root causes followed by an inference of the root cause of the current alarm flood. After the inference of the root cause, the alarm flood can be reduced to one alarm that caused the other alarms. This alarm is also the closest to the actual cause of the fault. However, this applies to a single alarm sequence. If multiple alarm sequences occur, there are also several root causes that are displayed, minimum one per alarm sequence.

In the following, a meaningful example is depicted to illustrate the effects of alarm floods. It also shows the importance to address this aspect in the field of research.

Example: Refinery Explosion

On the 24th July in 1994 an incident of the Pemproke Cracking Company Plant at the Texaco Refinery in Milford Haven happened. An explosion was followed by a number of fires caused by failures in management, equipment and control systems. These failures led to the release of 20 tonnes of flammable hydrocarbons from the outlet pipe of the flare knock-out drum of the fluidised catalytic cracking unit. These hydrocarbons caused subsequently the explosion. The failures started after plant disturbances appeared caused by a severe electrical storm.

The incident was investigated and analysed by the Health and Safety Executive (HSE) [11]. In total, there appeared 2040 alarms for the whole incident. 87% of these alarms were categorized in high priority. In the last 10.7 minutes before the explosion two operators had to handle 275 alarms. The alarm for the flare drum was activated approximately 25 minutes before the explosion but was not recognized by the operators. As a result, 26 persons were heavily wounded and the total cost of the economical damage was ca. €70 million. In a perfect scenario, where they would have found the crucial alarm immediately, the operators would have had 25 minutes to shut down the plant or at least minimize the possible damage caused by an explosion. This was impossible because of the flood of alarms so the operators could not handle the situation in an appropriate way.

Based on incidents like this, especially the chemical and oil industry expedited the topic of alarm management in industrial plants. One result is the guideline EEMUA 191 "Alarm Systems- A Guide to Design, Management and Procurement" by the non-profit organization Engineering Equipment & Materials Users' Association (EEMUA) [8]. The quasi-standard EEMUA 191 for alarm management recommends to have only one alarm per 10 minutes. The huge difference between this number and the 275 alarms from the example demonstrates the high potential for improvement. This is supported by the study of Bransby and Jenkinson [4]. They investigated 15 plants including oil refineries, chemical plants, pharmaceutical plants, gas terminal, and power stations. The average alarm rate per 10 minutes under normal operation ranged from 2 to 33 and the peak alarm rate per 10 minutes in plant upsets varied from 72 to 625.

Based on the guideline EEMUA 191, the International Society of Automation (ISA) developed a new standard called ANSI/ISA-18.2-2009 "Management of Alarm Systems for the Process Industries" in 2009 [17]. In 2014, another standard based on ISA 18.2 was developed by the International Electrotechnical Commission (IEC) the IEC62682:2014 [16]. The increasing focus of organizations and industry on the topic of alarm management shows the importance of reducing the amount of alarms in the future. In this challenge, the alarm flood reduction is one of the main tasks.

This is not only a beneficial effect for the safety of people, especially the employees, but also the plant itself. Moreover, the company can save a lot of money due to increased production and improved quality because the operator is able to focus better on the failures. This will also reduce the time to correct the failures and prevent unnecessary shut downs of parts of the plant.

In this work, we present an entire novel concept for alarm flood reduction in industrial plants. We depict the current status of alarm management in industrial plants and the state of the art in the field of alarm flood reduction (Section 2). For our approach a causal model, which represents the relations between the alarms in the plant is fundamental. Therefore, we discuss in Section 3 what knowledge is required for an accurate representation in the causal model. In our approach a Bayesian network is used as a causal model. In case of an alarm flood we are able to apply inference to identify the root cause and reduce the alarm flood to only the root cause. The whole approach is described in detail in Section 4. For the evaluation we applied our approach to a manufacturing plant in the SmartFactoryOWL. In the conclusion we give an outlook for further research which needs to be done to utilize the concept in real industrial plants.

2 State of the Art of Alarm Management

The basic intention of alarm management is to assist the operator in detecting, diagnosing and correcting the fault. Due to the advancing technology and automation, an increasingly number of alarms occur in plants, which requires a great amount of effort in the detection of faults. In addition, the risk of committing mistakes and unnecessarily exchanging parts increases. In the next paragraphs the current status of treating alarm floods in industrial plants is depicted and subsequently, relevant work in the field of alarm flood reduction is presented.

Current Status

The traditional practice for operators is using a chronologically sorted list-based alarm summary display [3]. During an alarm flood this alarm list reveals multiple weak points. In many cases the alarms occur faster than the human being is able to read them. The most common ways to handle alarm floods can be roughly grouped in one of the following approaches [5, 7, 26]:

- alarm shelving,
- alarm hiding/suppressing,
- alarm grouping,
- usage of priorities.

With alarm shelving the operator is able to postpone alarm problems till he has time to focus on the problem. This creates the opportunity to solve the problems subsequently. Alarm hiding means that some alarms are suppressed completely for special occasions like the starting procedure. Consequently, alarms that are expected, but irrelevant for this special situation, do not disturb the operator. Alarm grouping is used to create an alarm list which is clearer for the operator. Instead of many alarms, there will be only one alarm for one group. The technique of prioritizing enables the operator to identify immediately the most important alarms to prevent critical effects.

Most of these techniques just disguise the real problem. To reduce the amount of alarms in an alarm flood it is necessary to identify the real root cause or the alarm, which points out the real root cause. So that the operator is still provided with the required information to correct the faulty behaviour of the plant. In the next subsection some approaches for reducing the amount of alarms in an alarm flood are presented.

Related Work

There exist several concepts addressing the topic of alarm flood reduction. Wang et al. [28] give a comprehensive overview over the diverse ideas. For an overview of methodologies with probabilistic graphical models please refer to our previous work [29]. In this paragraph we will only focus on the most recent approaches in science. Most of them are related to the topic of pattern matching. Folmer et al. [9] developed an automatic alarm data analyzer (AADA) in order to identify the most frequent alarms and those causal alarms which allow to consolidate the alarm sequence. Ahmed et al. [2] use similarity analysis to investigate similar alarm floods in historical dataset. Based on the results they group patterns of alarms. A similar strategy is pursued by Fullen et al. [10]. They developed a case-based reasoning method on similar alarm floods. Cheng et al. [6] use a modified Smith-Waterman algorithm to calculate a similarity index of alarm floods by considering the time stamp information. Karoly and Abonyi [19] propose a multi-temporal sequence mining concept to extract patterns and formulate rules for alarm suppression. Xu et al. [30] introduce a data driven method for alarm flood pattern matching. With a modified BLAST algorithm using the Levenshtein distance they discover similar alarm floods. Rodrigo et al. [23] do a multiple steps causal analysis of alarm floods to reduce them. After removing chattering alarms and identifying alarm floods, they cluster similar alarm floods. Following this they try to isolate the causal alarm of an alarm flood. We want to focus on reducing alarm floods by identifying the root cause of it. Therefore we need a causal model which represents the dependencies of the alarms. Probabilistic graphical models, such as Bayesian nets, fault trees, or Petri nets are particularly suitable for this purpose. They have been already used in the field of alarm flood reduction. Especially Bayesian networks show great potential for this task.

Abele et al. [1] propose to combine modelling knowledge and machine learning knowledge to identify alarm root causes. They use a constrained-based method to learn the causal model of a plant represented by a Bayesian network. This enables faster modelling and accurate parametrization of alarm dependencies but expert knowledge is still required. Wang et al. [27] apply an online root-cause analysis of alarms in discrete Bayesian networks. They restrict the Bayesian network to have only one child node. The method is evaluated on a numerical example of a tank-level system. Wunderlich and Niggemann [29] investigate and evaluate different structure learning algorithms for Bayesian network in the field of alarm flood reduction. It turned out, that Bayesian networks are feasible and the state of the art structure learning algorithms are able to learn the causal relationships of alarms from industrial plants.

Based on the findings in our previous work, we developed a concept for the reduction of alarms. The foundation for this concept is the Bayesian network as a causal model. It is essential to learn a very accurate causal model in order to identify the causal chain and to determine the root cause. Therefore, it is important to know how a causal model can be learned in an unsupervised way and what information about the plant or process is necessary to be included.

3 Knowledge Representation

The research on alarm floods is still in an early stage and has a high potential of improvement. Therefore we want to find out, which knowledge is necessary and required to improve the situation. In general, once the amount of alarms is too high to be handled, the new arriving alarms will be ignored by the operator. This means the amount of alarms is randomly reduced which has a high potential of causing a disaster. Therefore, to improve the handling it is necessary to use all available knowledge to reduce the alarm flood in an intelligent way. Best case scenario would be if only the alarms which hint to the real causes, would be displayed. Therefore, it is a demanding task to identify all required causal relations between alarms [14]. To be able to achieve this, it is necessary to construct a causal model. Probabilistic graphical models are suitable for this use case. They represent in an easy-to-understand way the relationships and their probabilities. All these models are composed of nodes n and edges e. In our use case a node is equal to one alarm. The relations of two alarms (nodes) are represented by edges. For the causal model it is beneficial to include as much knowledge about the system as possible to be close to the reality. There are two extreme ways to achieve a representation of the real world such as phenomenological and first principle representation. The first approach is to learn the statistical relations of the alarms based on the alarm logs. However, it does not include every aspect of the plant. For example, as shown in Fig. 1, only the symptoms (alarms) are presented. But the alarms themselves may

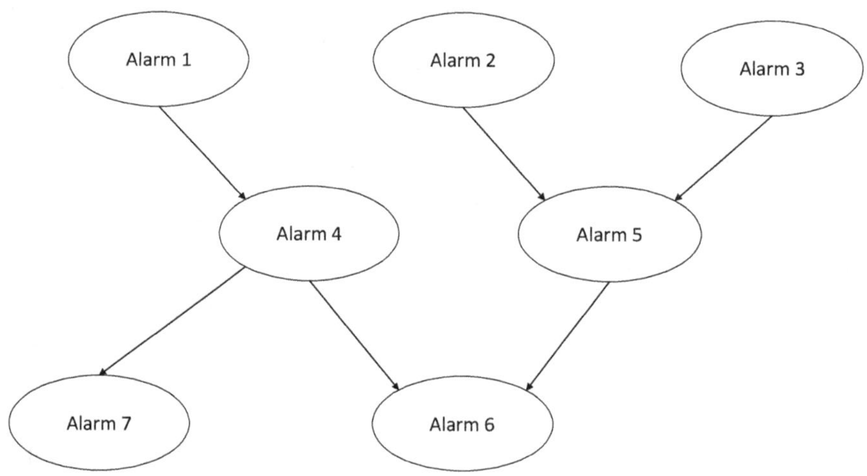

Fig. 1. Phenomenological representation

not be depended directly on each other. It's more viable to believe, that based on location or process flow the alarms are propagating and have effect on each other. The representation only depicts correlations and not causal relationships.

This leads to the second approach. Here, a first principle model of the system is built to create an exact image of the system. Not only the symptoms (alarms) are included, but also all aspects such as sensor values see Fig. 2. This has an advantage

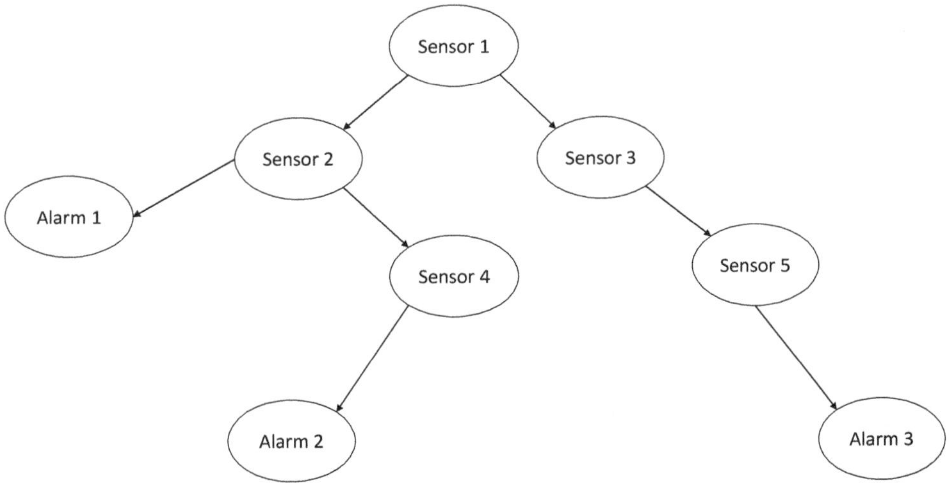

Fig. 2. First principle representation

of representing a deeper knowledge about the relations between the alarms and how the dependencies are. However, in most cases this is not possible due to missing information about some parts of the plant such as process flow. Also the creation of a model needs an expert and is very time-consuming. Hence, a combination of both is the most favourable solution. The model should include as much knowledge about the plant (process flow, environment etc.) as possible without being too time consuming. Finding a good balance is one of the challenges for researchers. In a best case scenario, the advantages of fast learning from historical data approach can be combined with the valuable expert knowledge about the plant and process.

We use a Bayesian network as a data-driven approach and try to fill it with as much information about the plant as possible.

4 Concept for Alarm Flood Reduction

Based on the findings, we design a concept for reducing alarm floods. To represent the causal relationships, we use Bayesian networks. An overview of the overall concept is shown in Fig. 3.

The concept consists of two different phases. First, the learning phase which is outlined in red and second, the operating phase which is outlined in green. In the learning phase, all historical data about the machine or system are used to learn a causal model of the alarms and their probabilities. This is done offline and usually takes several minutes to hours to calculate. The causal model contains information

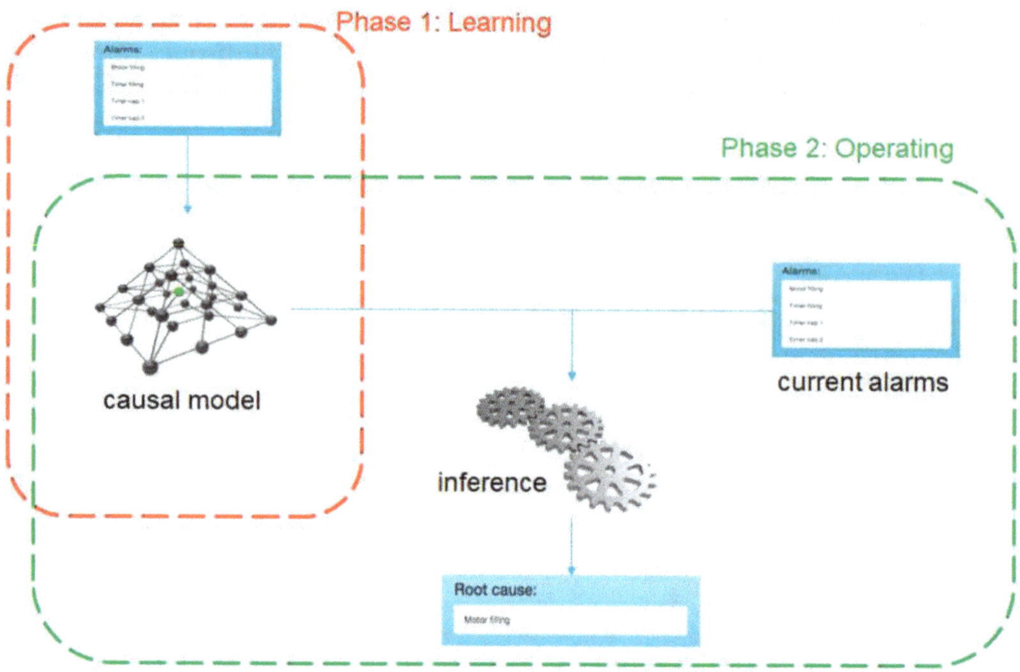

Fig. 3. Overview Concept Alarm Flood Reduction

about relations and dependencies of all alarms. However, the learning of the causal model is necessary only in the case of major changes of the system (for example, the production of a new product).

Once the learning of the causal model has been completed, the second phase, the so-called operating phase, can begin. The current alarms are used together with the causal model. With the aid of the causal model, it is possible to conclude the root cause of the current alarm sequence. This inference allows the number of alarms to be reduced to these root causes and thus to only display these root causes to the operator. Thus, the operator can correct the fault quickly and purposefully.

Demonstrator For the evaluation of this alarm flood reduction concept, we implemented two simple scenarios of alarm floods in the Versatile Production System (VPS), which is located in the SmartFactoryOWL [1]. This scenarios serve as use cases to identify the root cause. The VPS is a manufacturing plant in a laboratory scale where sensors, actuators, bus systems, automation components, and software from different manufacturers are considered. The VPS is a hybrid technical process considering both continuous and discrete process elements with a focus on the information processes and communication technologies from the plant level down to the sensor level. It thus provides an ideal multi vendor platform for testing and validation of innovative technologies and products.

[1] https://www.smartfactory-owl.de

The use cases are implemented in the "bottle filling module" of the VPS. The bottles can be filled either with water or with grain. The two use cases represent two possible root causes for an alarm flood. We want to identify these root causes with correctly learnt dependencies of alarms. In Fig. 4 an overview of the "bottle filling module" is depicted. The module consists of a conveyor belt and a rotary

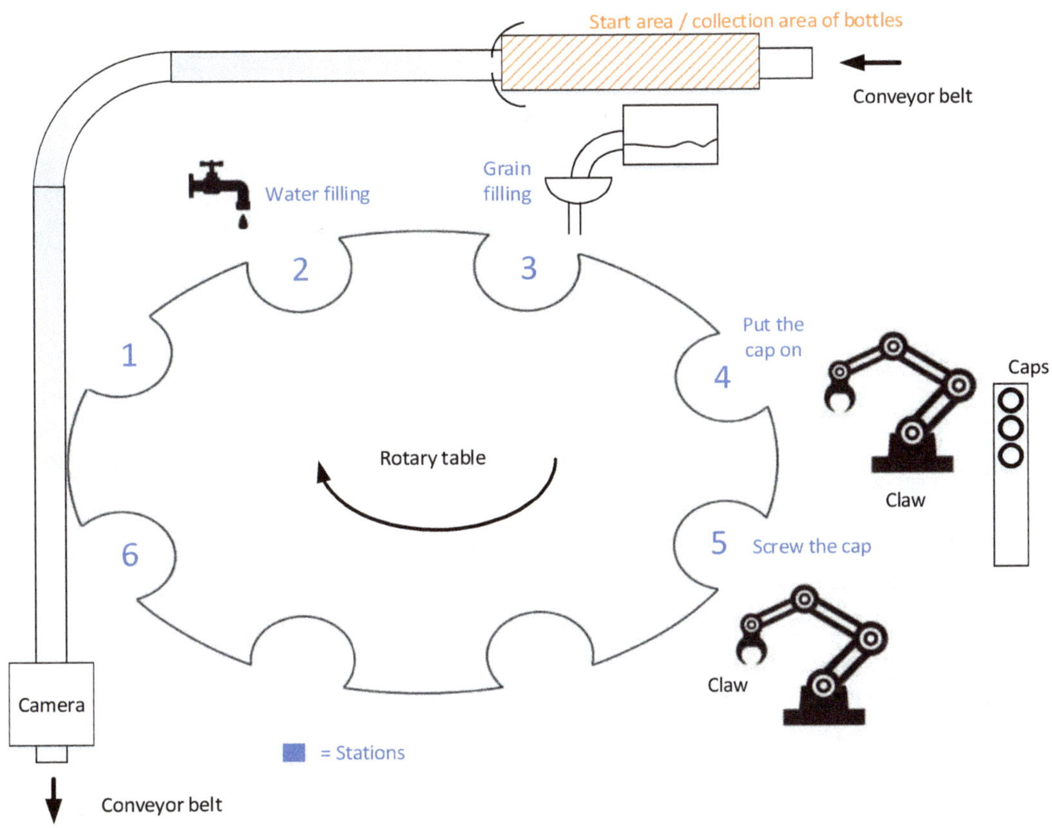

Fig. 4. Overview bottle filling module

table with six stations. At the first station the bottle is handed over by the conveyor belt to the rotary table. It's possible to fill the bottle either with water at station two or to fill it with grain at station three. The next two stations are there for putting the cap on. At the first step, the cap is placed on top of the bottle and in the second step, the cap is fastened. In the last station the bottle is handed over back to the conveyor belt passing a camera check, whether the bottle is filled.

In total the following alarms are implemented: bottle rotary table entrance (BRE), timer water tank (TW), drive filling (DF), timer filling (TF), timer cap 1 (TC1), timer cap 2 (TC2), bottle not filled (BNF). The real dependencies between the alarms are depicted in Fig. 5. This structure shows the real causality within the plant. The two root causes for an alarm flood are bottle rotary table entrance

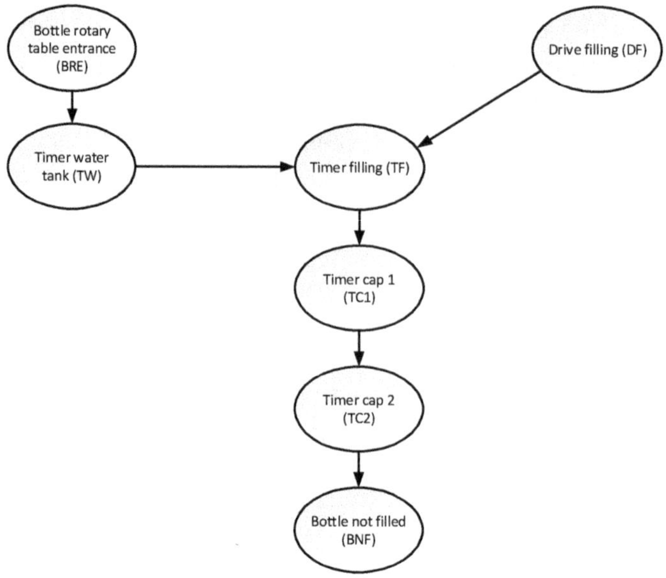

Fig. 5. Real causality structure

(BRE) and drive filling (DF). If a bottle is missing the alarm BRE triggers. This alarm causes the subsequent alarms TW, TF, TC1, TC2, and BNF. In the other root cause of a blocked drive the alarm DF causes the subsequent alarms TF, TC1, TC2, and BNF.

4.1 Learning Phase

Bayesian networks are a class of graphical models which allow an intuitive representation of multivariate data. A Bayesian network is a directed acyclic graph, denoted as $B = (N, E)$, with a set of variables $\boldsymbol{X} = \{X_1, X_2, \ldots, X_p\}$. Each node $n \in N$ is associated with one variable X_i. The edges $e \in E$, which connect the nodes, represent direct probabilistic dependencies.

Abele et al. [1] and Wang et al. [27] have already tried to use Bayesian networks for detection of a root-cause in an alarm flood. Abele et al. learned a Bayesian network, which is consequently the basis for root-cause analysis of a pressure tank system. The structure of the Bayesian network was learned with a constrained-based method and needed some expert knowledge to achieve the correct model of the pressure tank system. Therefore, they concluded that expert knowledge and machine learning should be combined for better results. Wang et al. applied a special kind of Bayesian networks. They restrict themselves to only one-child nodes. This is a huge restriction and it cuts off many possible failure cases, because in a modern and complex industrial plant the interconnectivity is increasing dramatically. This means that the alarms are also more connected and dependent on each other.

Structure Learning

Therefore, we want to pursue the idea of Abele et al. and use the Bayesian network as a model to represent the causality of the alarms. But other than Abele et al. we do not limit ourselves to constraint-based learning methods. All in all these learning algorithms can be differentiated into the following three groups of methods:

- constrained-based,
- score-based,
- hybrid.

In a constrained-based method the Bayesian network is a representation of independencies. The data is tested for conditional dependencies and independencies to identify a structure, which explains the dependencies and independencies of the data the best. Constrained-based methods are susceptible to failures in individual independence tests. Just one wrong answered independence test misleads to a wrong structure.

Score-based methods view Bayesian network as specifying a statistical model. Therefore, it is more like a model selection problem. In the first step, a hypothesis space of potential network structures is defined. In the second step, the potential structures are measured with a scoring function. The scoring function shows how good a potential structure fits the observed data. Following this, the computational task is to identify the highest-scoring structure. This task consists of a superexponential number of potential structures $2^{O(n^2)}$. Therefore, it is unsure if the highest-scoring structure can be found, so the algorithms use heuristic search techniques. Because the score-based methods consider the whole structure at once, they are less susceptible to individual failures and better at making compromises between the extent to which variables are dependent in the data and the cost of adding the edge [20].

Hybrid methods combine aspects of both constraint-based and score-based methods. They use conditional independence tests to reduce the search space and network scores to find the optimal network in the reduced space at the same time. In a previous work [29], we have already investigated different structural algorithms for reducing alarm flood. It turned out that a hybrid approach is the most promising due to the greater accuracy.

Algorithm

In the following, an hybrid method algorithm is presented and evaluated on the use cases. The Max-Min Hill-Climbing (MMHC) from Tsamardinos et al. [25] performed the best in previous tests and is chosen for the evaluation of the concept. MMHC is a hybrid of a constrained-based and score-based approach. In the first step, the search space for children and parents nodes is reduced by using a constrained-based approach, namely Max-Min Parents and Children (MMPC) algorithm. In the second step, the Hill-Climbing algorithm is applied to find the best fitting structure from the reduced search space.

For a better understanding of the associated pseudo code, we need a few definitions. The dataset D consists of a set of variables ϑ. In the variable PC_x the candidates of parents and children for the node X are stored. This set of candidates is calculated with MMPC algorithm. The variable Y is a node of the set PC_x. The pseudo code of MMHC looks as follows:

Algorithm 1 MMHC Algorithm

```
procedure MMHC(D)
    Input: data D
    Output: a DAG on the variables in D
    % Restrict
    for every variable X ∈ ϑ do
        PC_X = MMPC(X, D)
    end for
    % Search
    Starting from an empty graph perform Greedy Hill-Climbing with operators add-
  edge, delete-edge, reverse-edge. Y → X if Y ∈ PC_X
    Return the highest scoring DAG found
end procedure
```

The algorithm first identifies the parents and children set of each variable, then performs a greedy Hill-Climbing search in the reduced space of Bayesian network. The search begins with an empty graph. The edge addition, removal, or reversing which leads to the largest increase in the score is taken and the search continues in a similar way recursively. The difference from standard Hill-Climbing is that the search is constrained to only consider edges which where discovered by MMPC in the first phase. The MMPC algorithm calculates the correlation between the nodes. Based on a training dataset the MMHC algorithm gives the structure which is depicted in Fig. 6. The training data set was recorded at the VPS and consists of 525 observations of the seven alarms. The states of the alarms are binary coded with 0 for inactive and 1 for active.

The result is very close to the true causal model in Fig. 5. Both, the use-case with the missing bottle (BRE $\rightarrow$ TW $\rightarrow$ TF $\rightarrow$ TC1 $\rightarrow$ TC2 $\rightarrow$ FNB) and the use-case with the blocked drive (DF $\rightarrow$ TF $\rightarrow$ TC1 $\rightarrow$ TC2 $\rightarrow$ FNB) are shown correctly. Only the connection between DF and BRE is not present in reality.

In a graph with n nodes, there exist $n \cdot (n-1)$ possible connections. For the evaluation we define the following terms. A true positive connection (TP) is an edge which is in the original and in the learned Bayesian network. A false positive connection (FP) is an edge which is not in the original but in the learned Bayesian network. A false negative connection (FN) is an edge which is in the original but not in the learned Bayesian network. A true negative (TN) connection is an edge which is not in the original and not in the learned Bayesian network. The results of the evaluation for the MMHC algorithm is depicted in detail in Table 1.

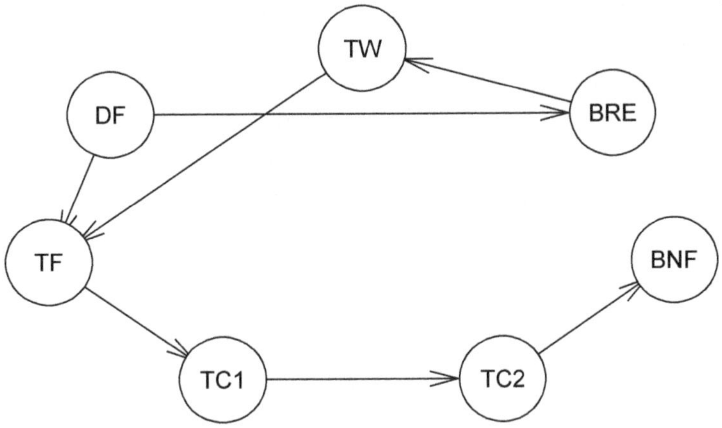

Fig. 6. BN Max-Min Hill-Climbing

Table 1. Results of evaluation

	MMHC
True Positive (TP)	6
False Positive (FP)	1
False Negative (FN)	0
True Negative (TN)	35
Accuracy [%]	97.62
F1-Score	0.92
Mean runtime [ms]	9.82

The MMHC learned all six edges from the original Bayesian network with the correct orientation. With only one misaligned edge (FP) and zero unwanted edges (FN) MMHC shows its strength. The accuracy is with 97.62% very good and underlined with an F1-score of 0.92. One slight disadvantage is the runtime. Based on the mean of 1000 runs the MMHC algorithm needs 9.82 ms which is quite long compared to other methods. However, it has to be considered, that in today's world the bottleneck is not the calculation power. The accuracy of the structure is more important. All in all, the hybrid method with the MMHC algorithm is best suited for learning a causal model from alarms. This was proven by Wunderlich and Niggemann [29], who evaluated different structure learning algorithms.

Parameter Learning

Once the structure of the causal model has been learned, it is still necessary to calculate or estimate the probabilities for the dependencies. Only in the combination of structure and parameters (probabilities) the inference of the root cause can be done in the operation phase. To learn the parameters, the classical method maximum likelihood estimation (MLE), which was developed by R.A. Fischer, is used. Here, a parameter p is estimated to maximize the probability of obtaining the observation under the condition of the parameter p. In other words, the MLE provides the most plausible parameter p as an estimate with respect to the observation. If the parameter p is a probability in the Bayesian network and the historical data D represents the observations, the likelihood function is composed as follows:

$$L(D|p) = \prod_{i=1}^{n} f(D|p) \tag{1}$$

The probability density function of D under the condition p is $f(D|p)$. In Figure 7 the causal model with the learned probabilities is shown. It is noteworthy that the

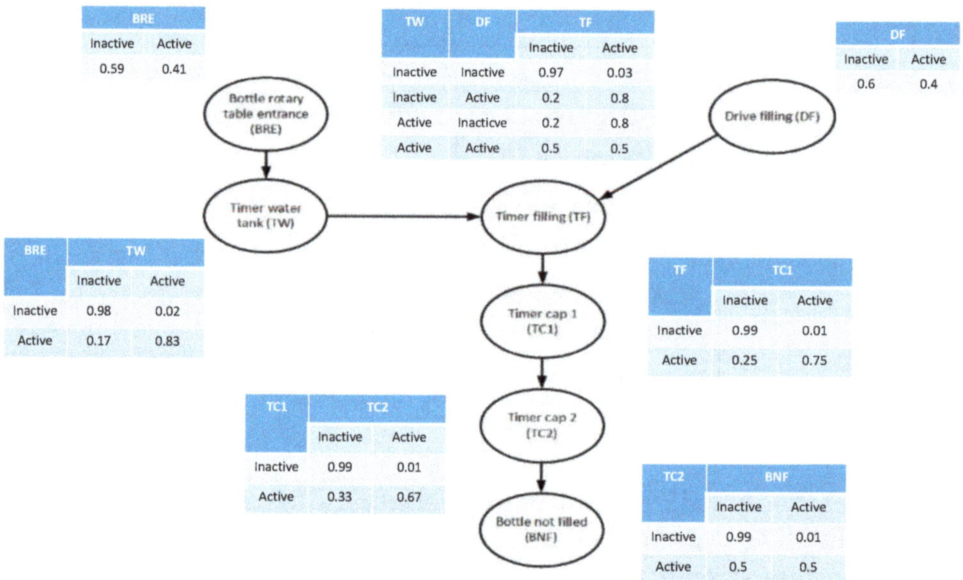

Fig. 7. Bayesian net with Probabilities

variant with active alarms TW and DF never occurs so there are no probabilities in this regard. With the learned structure and the probabilities, the inference can begin.

4.2 Operation Phase

There are two different variants for the inference, namely the exact inference and the approximate inference. For the exact inference, the probabilities are calculated specifically for the query. One famous method of exact inference is the variable elimination (VE). In doing so, variables irrelevant to the query are eliminated from

the probability distribution. This is computationally very complex and expensive. Therefore, such a method is only feasible for very small Bayesian networks.
An alternative and more frequently used are the approximate inference methods. Here the model, which is represented by the Bayesian network, is randomly simulated. This process is called sampling. This makes it possible to approximate the probability of the query. For example, it can be determined how high the probability is that a particular node assumes a specific state. The cases in which this is occurred are counted and set in proportion to the total sample size. A disadvantage of this method is that under certain circumstances, a large number of samples are required in order to provide a reliable result and thus significantly increase the calculation time.
To evaluate our concept, we opted for the simple and fast logic sampling (LS) algorithm. The LS algorithm is a very simple procedure developed by Max Henrion in 1986 [12]. In this case, a state is arbitrarily assumed per sample for the root nodes according to their probability table. Thus, a certain number of samples, which are determined, are carried out. Subsequently, the probability that e.g. a node X assumes the state True as follows:

$$P(X = True) = \frac{\text{No. of cases with X=True}}{\text{No. of all samples}} \tag{2}$$

This process always converges to the correct solution, but in very rare cases the number of samples required can become exorbitant [18].

Inference

In this subsection we apply and evaluate the inference with the LS algorithm on the VPS demonstrator. In the previous subsection it was shown, that a decent structure could be learned. Combined with time-limited expert knowledge we achieved an accurate causal model. Based on this causal model and the learned probabilities the inference of the root causes is enabled. For the inference of an alarm flood in the demonstrator we use as possible root causes the two alarms BRE and DF. This is given by the learned structure of the Bayesian network. For the evaluation we investigate the use case of a missing bottle. Therefore we have as evidence E the alarms (TW, TF, TC1, TC2, BNF) and formulate the following two queries.

$$P(BRE = active|E = active) = \frac{\text{No. of cases with BRE=active given E=active}}{\text{No. of all samples}} \tag{3}$$

$$P(DF = active|E = active) = \frac{\text{No. of cases with DF=active given E=active}}{\text{No. of all samples}} \tag{4}$$

The two queries calculate the probability of BRE or DF to be active given that all alarms of evidence are active. In our use case, the estimation shows a probability of 97 % for BRE to be active and a probability of 40 % for DF to be active. This results in the conclusion, that the alarm BRE is the root cause for the current alarm flood.

A comparison of the concept for the use case of a missing bottle in the VPS is shown in Table 2.

Table 2. Alarm Flood Reduction

Without	With
Timer Filing	Bottle Rotary Table Entrance
Timer Cap 1	
Timer Cap 2	
Bottle Not Filled	
Bottle Rotary Table Entrance	
Timer Water Tank	

If a bottle is missing, the six alarms (Timer Filing, Timer Cap 1, Timer Cap 2, Bottle Not Filled, Bottle Rotary Table Entrance and Timer Water Tank) will appear. The operator is not immediately able to identify the cause of this alarm sequence. It changes when our concept is applied in this scenario. The amount of alarms can be reduced from six to only one alarm. This one alarm BRE is also the cause of the alarm flood. This result allows the operator to quickly and efficiently identify the cause and take the necessary steps to remedy the problem. A similar result is also obtained for the other use case of a blocked drive for grain filling.

5 Conclusion

We presented the increasing problem of overwhelming alarm floods in industrial plants. One way to solve this problem is to reduce the alarm floods, especially sequences of alarms caused by one alarm. Therefore, we propose a concept to identify the real root cause of an alarm flood using Bayesian networks. The Bayesian network serves as a causal model and enables inference about the root cause. Instead of all alarms, only the root cause is depicted to the operator. This supports the operator to take better care of the plant. The concept is evaluated on real use cases of a demonstrator in the SmartFactoryOWL. For the use cases the concept shows a promising result identifying the real root cause. Obviously, the demonstrator is still quite small and not complex compared to real industrial plants. Therefore, it is necessary to evaluate how this approach scales on real industrial plants, where also additional challenges appear. This not only means the increasing complexity, but also the occurrence of disturbances or missing or incorrect historical data records.

Nevertheless, we think that Bayesian networks are a good foundation to learn a causal model of the dependencies of alarms in a plant. Bayesian networks are robust against uncertainty such as incomplete or defective data. Because it's a data-driven approach it reduces the amount of time in which an expert is needed for constructing the causal model. The expert knowledge is still necessary, but the approach with Bayesian network can be improved by including time behaviour or dynamic process like a product change in the plant. Also the algorithms for learning the structure can be improved by using methods like Transfer Entropy for a better edge orientation.

Acknowledgment

This work was supported by the German Federal Ministry of Education and Research (BMBF) under the project ADIMA (grant number: 13FH019PX5) and by the European Union's Horizon 2020 research and innovation programme under the project IMPROVE (grant agreement No. 678867).

References

1. Abele, L., Anic, M., Gutmann, T., Folmer, J., Kleinsteuber, M., Vogel-Heuser, B.: Combining knowledge modeling and machine learning for alarm root cause analysis. IFAC Proceedings Volumes 46(9), 1843 – 1848 (2013)
2. Ahmed, K., Izadi, I., Chen, T., Joe, D., Burton, T.: Similarity analysis of industrial alarm flood data. IEEE Transactions on Automation Science and Engineering 10(2), 452–457 (April 2013)
3. Bransby, M.: Design of alarm systems. IEE CONTROL ENGINEERING SERIES pp. 207–221 (2001)
4. Bransby, M., Jenkinson, J.: The management of alarm systems: a review of best practice in the procurement, design and management of alarm systems in the chemical and power industries. Technical Report CRR 166, Health and Safety Executive (1998)
5. Bullemer, P.T., Tolsma, M., Reising, D., Laberge, J.: Towards improving operator alarm flood responses: Alternative alarm presentation techniques. Abnormal Situation Management Consortium (2011)
6. Cheng, Y., Izadi, I., Chen, T.: Pattern matching of alarm flood sequences by a modified smithwaterman algorithm. Chemical Engineering Research and Design 91(6), 1085 – 1094 (2013)
7. Directorate, Norwegian Petroleum: Principles for alarm system design. YA-711, Feb (2001)
8. Engineering Equipment and Materials Users' Association: Alarm Systems: A Guide to Design, Management and Procurement. EEMUA publication, EEMUA (Engineering Equipment & Materials Users Association) (2007)
9. Folmer, J., Vogel-Heuser, B.: Computing dependent industrial alarms for alarm flood reduction. Transactions on Systems, Signals and Devices (TSSD) PP (2013)
10. Fullen, M., Schüller, P., Niggemann, O.: Semi-supervised case-based reasoning approach to alarm flood analysis. In: 3rd Conference on Machine Learning for Cyber Physical Systems and Industry 4.0 (ML4CPS) (10 2017)

11. Health and Safety Executive: The Explosion and Fires at the Texaco Refinery, Milford Haven, 24 July 1994: A Report of the Investigation by the Health and Safety Executive Into the Explosion and Fires on the Pembroke Cracking Company Plant at the Texaco Refinery, Milford Haven on 24 July 1994. Incident Report Series, HSE Books (1997)
12. Henrion, M.: Propagating uncertainty in Bayesian networks by probabilistic logic sampling. In: Uncertainty in Artificial Intelligence 2 Annual Conference on Uncertainty in Artificial Intelligence (UAI-86). pp. 149–163. Elsevier Science, Amsterdam, NL (1986)
13. Hollender, M., Skovholt, T.C., Evans, J.: Holistic alarm management throughout the plant lifecycle. In: 2016 Petroleum and Chemical Industry Conference Europe (PCIC Europe). pp. 1–6 (June 2016)
14. Hollender, M., Beuthel, C.: Intelligent alarming. ABB review 1, 20–23 (2007)
15. Hollifield, B., Habibi, E., Pinto, J.: The Alarm Management Handbook: A Comprehensive Guide. PAS (2010)
16. International Electrotechnical Commission: EN-IEC 62682:2014 Management of alarm systems for the process industries. Tech. rep., International Electrotechnical Commission
17. International Society of Automation and American National Standards Institute: ANSI/ISA-18.2-2009, Management of Alarm Systems for the Process Industries. ISA (2009)
18. Kanal, L., Lemmer, J.: Uncertainty in Artificial Intelligence 2. Machine Intelligence and Pattern Recognition, Elsevier Science (2014)
19. Karoly, R., Abonyi, J.: Multi-temporal sequential pattern mining based improvement of alarm management systems. In: Systems, Man, and Cybernetics (SMC), 2016 IEEE International Conference on. pp. 003870–003875. IEEE (2016)
20. Koller, D., Friedman, N.: Probabilistic Graphical Models: Principles and Techniques - Adaptive Computation and Machine Learning. The MIT Press (2009)
21. Larsson, J.E., Öhman, B., Calzada, A., DeBor, J.: New solutions for alarm problems. In: Proceedings of the 5th International Topical Meeting on Nuclear Plant Instrumentation, Controls, and Human Interface Technology, Albuquerque, New Mexico (2006)
22. Pearl, J., Bacchus, F., Spirtes, P., Glymour, C., Scheines, R.: Probabilistic reasoning in intelligent systems: Networks of plausible inference. Synthese 104(1), 161–176 (1995)
23. Rodrigo, V., Chioua, M., Hägglund, T., Hollender, M.: Causal analysis for alarm flood reduction. In: 11th IFAC Symposium on Dynamics and Control of Process SystemsIncluding Biosystems. IFAC (2016)
24. Rothenberg, D.: Alarm Management for Process Control: A Best-practice Guide for Design, Implementation, and Use of Industrial Alarm Systems. Momentum Press (2009)
25. Tsamardinos, I., Brown, L.E., Aliferis, C.F.: The max-min hill-climbing Bayesian network structure learning algorithm. Machine Learning 65(1), 31–78 (2006)
26. Urban, P., Landryov, L.: Identification and evaluation of alarm logs from the alarm management system. In: 2016 17th International Carpathian Control Conference (ICCC). pp. 769–774 (May 2016)
27. Wang, J., Xu, J., Zhu, D.: Online root-cause analysis of alarms in discrete bayesian networks with known structures. In: Proceeding of the 11th World Congress on Intelligent Control and Automation. pp. 467–472 (June 2014)
28. Wang, J., Yang, F., Chen, T., Shah, S.L.: An overview of industrial alarm systems: Main causes for alarm overloading, research status, and open problems. IEEE Transactions on Automation Science and Engineering 13(2), 1045–1061 (April 2016)

29. Wunderlich, P., Niggemann, O.: Structure learning methods for Bayesian networks to reduce alarm floods by identifying the root cause. In: 22nd IEEE International Conference on Emerging Technologies and Factory Automation (ETFA 2017). Limassol, Cyprus (Sep 2017)
30. Xu, Y., Tan, W., Li, T.: An alarm flood pattern matching algorithm based on modified BLAST with Leveshtein distance. In: Control, Automation, Robotics and Vision (ICARCV), 2016 14th International Conference on. pp. 1–6. IEEE (2016)

www.ingramcontent.com/pod-product-compliance
Lightning Source LLC
LaVergne TN
LVHW082237200926
848484LV00008B/2370